INTRODUCTION TO SUPERSYMMETRY AND SUPERGRAVITY

Introduction to Supersymmetry and Supergravity

S.P. MISRA
Professor, Institute of Physics
Bhubaneswar, India

JOHN WILEY & SONS
NEW YORK CHICHESTER BRISBANE TORONTO SINGAPORE

First Published in 1992 by
WILEY EASTERN LIMITED
4835/24 Ansari Road, Daryaganj
New Delhi 110 002, India

Distributors:

Australia and New Zealand:
JACARANDA WILEY LIMITED
PO Box 1226, Milton Old 4064, Australia

Canada:
JOHN WILEY & SONS CANADA LIMITED
22 Worcester Road, Rexdale, Ontario, Canada

Europe and Africa:
JOHN WILEY & SONS LIMITED
Baffins Lane, Chichester, West Sussex, England

South East Asia:
JOHN WILEY & SONS (PTE) LIMITED
05-04, Block B, Union Industrial Building
37 Jalan Pemimpin, Singapore 2057

Africa and South Asia:
WILEY EASTERN LIMITED
4835/24 Ansari Road, Daryaganj
New Delhi 110 002, India

North and South America and rest of the World:
JOHN WILEY & SONS, INC.
605 Third Avenue, New York, NY 10158, USA

Copyright © 1992 WILEY EASTERN LIMITED
New Delhi, India

Library of Congress Cataloging-in-Publication Data

ISBN 0-470-21862-2 John Wiley & Sons, Inc.
ISBN 81-224-0409-X Wiley Eastern Limited

Printed in India at Baba Barkha Nath Printers, New Delhi

To My Wife

Foreword

The Science and Engineering Research Council (SERC) is a high-level body which advises the Department of Science and Technology on matters concerning frontier research and development activities in various science and engineering disciplines.

SERC has adopted a two-pronged strategy for fulfilling its objectives. First, it supports time-bound R & D projects which are carefully refereed and processed through a peer review mechanism. Second, in order to meet the long-term manpower requirements, SERC has also initiated a major programme of organising summer/winter schools in selected thrust areas to encourage young scientists to take up research at various academic institutions and laboratories in the country. Under the programme, several series of SERC schools are being organised in various identified thrust areas like Theoretical High Energy Physics, Condensed Matter Physics, Nuclear Physics, etc. It is proposed that in each of these areas, there should be a cycle of such schools - each cycle consisting of about five schools. The primary concern here is to be able to assess the impact of these schools on manpower development. Isolated workshops may not be expected to lead to any such impact.

One of the first series to be organised was in the area of Theoretical High Energy Physics (THEP). The series of five schools was started in 1985. The techniques and ideas in this frontline area, which aim at learning about the elementary constituents matter and their interactions have been changing very fast. Realising continuing need for young and energtic manpower, suitably trained to address contemporary problems, the Indian high energy physics community took lot of interest in this exercise.

It was felt that the quality, nature and relevance of the lectures delivered at these schools were such that the lectures, if published, would be useful for a much wider readership (even at the international level). Also, if made available to students and research community in India at an affordable price, it would further aid in manpower development, a primary objective of the school. It was, however, decided that the lectures be suitably written, reorganised (if required, not necessarily schoolwise) and published in the form of books/monographs having the requiste pedagogy and continuity to form as material for training young and aspiring researchers. Conventional conference proceedings do not have this character. The present monograph has come out of the first five year cycle of SERC schools mentioned above. About ten monographs are expected to come out of this series. It is hoped that this effort will be useful for high energy physicists, in general, and young

Indian researchers in this area, in particular.

On behalf of the Department of Science and Technology, I would like to particularly thank Professor N. Mukunda for taking the difficult responsibility of being the Series Editor of these monographs. The authors of individual monographs also deserve special thanks for timely completion of their manuscripts with meticulous care and proficiency.

New Delhi
September 17, 1991

P. Rama Rao
Chairman
Science and Engineering
Research Council

Editor's Note

This monograph is one of a series devoted to topics in theoretical high energy physics, originating from courses given at the Schools in this subject being supported on a continuing basis by the Science and Engineering Research Council of the Department of Science and Technology, Government of India. These Schools are being held approximately once a year since 1985. They are carefully planned and conducted by high energy physics community in India as a collaborative effort. Considerable time and care are devoted to the choice of courses and preparation of material to be presented at each School.

The audience at which these Schools are aimed consists basically of graduate students in high energy physics working towards the doctoral degree in various institutions in the country. In addition, some students in allied fields like nuclear physics, relativity, astrophysics and cosmology as well as some more senior and experienced research workers with interest in specific topics, are also included. The courses given at each School have both a strong pedagogical component, conveyed through major long courses; and a topical component covering recent and exciting advances through shorter courses. Attention is given to development of basic theoretical techniques, related useful mathematical material, and phenomenology.

Each monograph in this series puts together courses from one or more Schools grouped around a common theme, in a way which is likely to be most useful to the reader, and makes the basic material available in a compact form. We expect these monographs will remain useful pedagogical sources of information for fresh entrants into the field of high energy physics, and to others.

The present monograph is a self-contained and carefully written introduction to the topics of supersymmetry and supergravity. It originated in a major course of lectures on these topics at the 1986 School, later supplemented by further lectures of the author, S.P. Misra, at the Institute of Physics, Bhubaneswar. Considering the extent of the material covered, it was felt desirable to devote an entire monograph to it, and not combine it with other material presented by other lecturers. We are grateful to Professor S.P. Misra for the special effort put in by him in making these notes especially readable, assimilable and complete for a beginner. His own preface explains in detail the special points of view he has adopted in his presentation.

N.Mukunda

Preface

The present book arises from a number of lectures in a Summer School organised by the Department of Science and Technology, Government of India on Supersymmetry and Supergravity for researchers in High Energy Physics as well as lectures delivered to the Ph.D. students at the Institute of Physics, Bhubaneswar. Since there are now many books on both the subjects dealing with pedagogical aspects of the same, this aspect has not been dealt with in detail here. What has been aimed at is the coverage of a large sector of the interrelated fields in one book, keeping the volume within limits so that the students in the subject may not get lost in one aspect of the theory while being unaware of its many other interesting facets. We thought that emphasizing this for supersymmetry and supergravity is particularly important since the subject at present is such that which will become relevant in future can not be predicted. The subject itself from both mathematics and physics points of view has an extremely wide spread.

In the present book, we also give emphasis to grand unification type of phenomenology because such an unification is inevitable, and we believe that supersymmetry and supergravity type of approach is the most likely candidate for the same. When this happens, the conceptual as well as symmetry aspects of the same will become necessary both for developing or even understanding it. With this in mind we think it desirable to emphasize some aspects, wherein new type of openings or ideas could be relevant. This naturally has a subjective side to it, which will be obvious in the materials presented in the book. While illustrating the principles as examples, more of the work in which the author is directly involved, has been presented. This is just because the author found it easier to illustrate the points of view with the examples considered by him than with the corresponding examples considered by other people. It was natural while giving lectures, and has persisted here. In this sense, the book does not attempt to be a complete review of the subject as it was thought that such a task, in fact, will be horrendous and even may not be worthwhile. What the author, however, does regret is the absence of a reasonably complete reference. This appeared to be almost impossible as the subject matter has a lot of activity and changes direction much too rapidly. The emphasis, therefore, has been on diversity rather than detail.

The objective of the book has been, while giving some basic information, to arouse the curiosity of the students and make them realise the diverse ways in which high energy physics might develop in future. If this purpose is even partly fulfilled, the author will consider himself amply rewarded.

In a complicated subject like supergravity, there are bound to be a number of trivial as well as non-trivial errors. It is not attempted to eliminate all of them, since the physics output is mostly qualitative, and a possible error of factors does not change the picture. However, the concepts with or without superstring bias, are fascinating in their generality and depth, both in the context of physics application as well as sophispticated mathematical ideas which have become a part of our thinking in physics. The book should help in developing familiarity, appreciation and understanding regarding the same on the part of beginners, and that has been our main consideration here.

I would like to thank Professor N.Mukunda and Dr.P.J.Lavakare as well as Department of Science and Technology, Government of India for their keen interest regarding the writing of the book, to Dr. A.R.Panda, Dr. Snigdha Mishra, H.Mishra, S.N.Nayak, A.Mishra and P.K.Panda and others for discussions, and to many students who responded with curiousity and excitement during the lectures and shared enthusiasm and insight with me. I am also thankful to Anjan and Rukmani for carefully reading the manuscript.

July 1991   **S.P. Misra**

Contents

Foreword   vii

Editor's Note   ix

Preface   xi

1. **Introduction**   1

2. **Supersymmetry in Quantum Mechanics**   7
   - 2.1 Grassmann variables   7
   - 2.2 Superspace in Quantum Mechanics   7
   - 2.3 Supersymmetry Transformation   8
   - 2.4 Invariant Lagrangian   9
   - 2.5 Action Principle in Superspace   10
   - 2.5.1 Equations of Motion   11
   - 2.5.2 Quantisation   12
   - 2.5.3 Hamiltonian   13
   - 2.5.4 Supersymmetric Charges   14
   - 2.6 Remarks   15

3. **SL(2,C) Representations of the Lorentz Group**   17
   - 3.1 SL(2,C) and 2-spinors   17
   - 3.2 Dirac Spinors   20

4. **Superfields**   23
   - 4.1 Introduction   23
   - 4.2 Chiral Superfields   25
   - 4.3 Vector Superfields   26
   - 4.4 Supersymmetric Lagrangian   28
   - 4.4.1 Chiral Superfields   28
   - 4.4.2 Vector Superfields   32
   - 4.5 Outlook   35

5. **Local Gauge Symmetry**   37
   - 5.1 Introduction   37
   - 5.2 Gauge Invariant Lagrangian   37
   - 5.2.1 Lagrangian for Matter Field   38
   - 5.2.2 Lagrangian for Gauge Field   41

5.3	Electroweak Theory	46

6. Symmetry Breaking — 52
- 6.1 Introduction — 52
- 6.2 O'Raifeartaigh Mechanism — 53
- 6.3 Fayet-Illiopoulos Mechanism — 54
- 6.4 Supersymmetric SU(5) Model — 57
- 6.5 Low Energy Sector — 59
- 6.6 Quantum Phase Transition — 61
- 6.7 Grand Unification Models — 64
- 6.7.1 SU(5) — 64
- 6.7.2 SO(10) — 65
- 6.7.3 E_6 and E_8 — 65
- 6.7.4 Proton decay, R-symetry and R-parity — 66

7. Functional Methods in Superspace — 69
- 7.1 Integration of GrassmannVariables — 69
- 7.2 Superspace Dynamics — 70
- 7.3 Cancellation of Divergences — 76

8. Superspace and Supergravity: A Geometric Picture — 81
- 8.1 Introduction — 81
- 8.2 Differential Forms in Superspace — 81
- 8.3 Local Supersymmetry — 84
- 8.4 Gravity Multiplet — 88
- 8.5 Chiral and Vector Multiplets — 88
- 8.6 Chiral Densities and the Lagrangian — 90
- 8.6.1 Chiral Densities — 91
- 8.6.2 Lagrangian Construction — 91

9. Supergravity Lagrangian — 93
- 9.1 Introduction — 93
- 9.2 Lagrangian Construction — 94
- 9.3 Identification of the Terms — 98

10. Model Building in Supergravity — 103
- 10.1 Introduction — 103
- 10.2 SU(5): — 104
- 10.3 SO(10) Models — 107
- 10.3.1 Single Stage Symmetry Breaking — 107
- 10.3.2 Models with Intermediate Scales — 112
- 10.3.3 Generation of Unification Hierarchies — 113
- 10.4 Superheavy Higgs Particles — 118

10.5	Radiative Symmetry Breaking	122
10.6	Remarks	125

11. Supergravity in Ten Dimensions — 128
11.1	Introduction	128
11.2	Supergravity Lagrangian in Ten Dimensions	128
11.3	Heterotic Superstring Theory	133
11.4	Discussions	138

12. Compactification of Higher Dimensional Theories — 139
12.1	Introduction	139
12.2	Compactification in 10 Dimensions	139
12.2.1	Masses in Four Dimensions	140
12.2.2	Witten's Prescription	144
12.2.3	Geometry of Extra Dimensions	146
12.3	Superstring Based Supergravity	152
12.3.1	Wilson Loops	153
12.3.2	Spaces with $Z_5 \times Z_5$ Discrete Symmetry	156
12.4	Three Generation Models	157
12.4.1	Symmetry after Compactification	158
12.4.2	Yukawa Couplings and Symmetries	160
12.4.3	Intermediate Scales	165
12.4.4	Compactification with E_6 Going to $SU(6) \times U(1)$	169
12.5	Discussions	170

13. Supersymmetry in Two Dimensions — 175
13.1	Introduction	175
13.1.1	Spinors in 1+1 Dimensions	175
13.1.2	Superspace for 1+1 Dimensions	176
13.2	Supersymmetry in 1+1 Dimensions	177
13.2.1	Superfields in 1+1 Dimensions	177
13.2.2	Lagrangian Construction	178
13.2.3	Supersymmetry	179
13.2.4	Field expansions	182
13.3	Quantum Phase Transition	183
13.3.1	Breaking of Supersymmetry	184
13.3.2	Symmetry Breaking at Zero Temperature	184
13.4	Temperature Dependance of Phase Transition	186
13.5	Discussions	192

14. Conclusions — 193
14.1	Experimental Signatures	193

14.2	Preonic Models	195
14.3	Matrix Models	200
14.4	Remarks	204

Appendix A: Construction of Calabi Yau Manifolds — 207

A.1	Manifolds	207
A.2	Complex Structures	208
A.3	Hermitian Complex Manifolds	208
A.4	Differential Forms	209
A.5	Connections and Curvature	210
A.6	Kahler Manifold	210
A.7	Calabi-Yau Manifold	211
A.8	Betti Hodge Numbers	213
A.9	Examples of Calabi Yau Space	213
A.10	Examples of CICY spaces	215
A.10.1	CHSW Manifold	215
A.10.2	Tian Yau Manifold	216
A.11	Remarks	217
A.11.1	Harmonic Forms	217
A.11.2	Chern Classes and Euler Characteristic	218
A.11.3	Betti-Hodge Numbers and Complex Deformations	220

Appendix B: Some Results in Group Theory — 224

B.1	Dynkin Diagrams	224
B.2	Dynkin Labels	225
B.3	Representations	226
B.4	Symmetry Breaking	229
B.5	E_6 Symmetry Group	230
B.5.1	Embedding of G_{321} in E_6	231

Index — 237

Chapter 1

Introduction

Supersymmetry consists of transformations of fermions to bosons and vice versa. This is usually associated with the Lorentz group since in four dimensions it gets associated with spin. Thus it really is a space-time symmetry as opposed to internal symmetries. We shall consider the theory at an elementary level, but include in our discussions a wide class of concepts and techniques which have got associated with supersymmetry and supergravity (or local supersymmetry).

Let us first recognise what is meant by fermions and bosons. Fermions are defined with mathematical objects which on quantisation are associated with an anticommutator algebra. Thus the corresponding many body systems will have exclusion principle and Fermi statistics. Similarly bosons are defined with objects where quantisation is associated with commutator algebra, which thus leads to Bose statistics. Generally these need not be associated with spin. E.g. for quantum mechanics, the concept of spin does not exist and still we can have fermions and bosons.

We have considered this ambiguity in defining what we mean by fermions and bosons because we shall consider supersymmetry in arbitrary dimensions. Thus, in the next Chapter itself, we shall discuss supersymmetric quantum mechanics which is supersymmetry in (1,0) dimensions. The fermion degree of freedom here consists of two distinct sectors with either zero fermion number or with fermion number one. This is a simplifying feature for the consideration of supersymmetric quantum mechanics. Thus a state will be fermionic if the fermion number is one and will be bosonic if the fermion number is zero. We take a format for the description of the same so that it leads to supersymmetry in four dimensions.

In Chapter 3 we give a very brief account of the representations of the Lorentz group. This is essential since these representations in four dimensions yield fermions or bosons, and supersymmetric transformations will have to connect the different representations of the Lorentz group. In particular here we define Weyl spinors, Majorana spinors and Dirac spinors to be used subsequently for the definition of supersymmetric transformations, where the "increments" will be necessarily spinorial in nature enabling to link fermions with bosons.

In Chapter 4 we discuss chiral superfields, vector superfields and the construction of the supersymmetric Lagrangians. This is tackled through the definition of superfields on a superspace. A superspace is a generalisation of ordinary space where we introduce fresh dimensions with additional fermionic coordinates. This is a mathematical technique which enables us to describe fermion-boson transformations in a reasonably simple manner, and helps us in constructing Lagrangians which are symmetric under these transformations. In Chapter 5 we generalise these techniques to include nonabelian gauge theories. These are essential inputs for the unification of the laws of physics, as e.g. Salam Weinberg theory for electroweak unification. We also know that such nonabelian gauge theories are broken at low energies. Such spontaneous symmetry breaking is discussed in Chapter 6. Here we note that at low energies there is not only a symmetry breaking of the nonabelian gauge theories but also a breaking of supersymmetry itself. The mechanism for the breaking of supersymmetry as a global symmetry is discussed with some examples. An approach to symmetry breaking as vacuum instability in quantum field theory is also discussed.

We have said that supersymmetry is associated with fermion-boson symmetry and thus the increments must be fermionic in nature. We introduce *superspace* with fermionic coordinates so that translations in such coordinates can be the increments for supersymmetric transformations. Superspace however introduces a fresh conceptual feature in the description of the path integral method in field theory. We introduce such techniques briefly in Chapter 7. The cancellation of divergences in a superspace formulation of the underlying dynamics between bosonic and fermionic degrees of freedom becomes relatively simple, whereas in the conventional field theory picture with separate fermionic and bosonic component fields the above cancellation looks "magical". We illustrate this through the nonrenormalisation theorem in superspace.

We devote Chapter 8 to the description of supergravity, or local supersymmetry, through a geometric picture. For this purpose we introduce the appropriate differential forms in superspace. We first discuss ordinary global supersymmetry in superspace, where, we notice that the space is flat, but it has a nonvanishing torsion in superspace which are given by the Dirac matrices. We now include local coordinate transformations in superspace, along with the idea of curvature and torsion in superspace. We then note that the framework has far too many fields to be sensible as a physical model for the world as we know it. This forces us to introduce appropriate constraints in a selfconsistent manner. This is done with repeated use of Bianchi identities in superspace. An interplay of local coordinate transformations and Lorentz transfromations then enables us to construct a reasonably minimal description. The local supersymmetric transformations for appropriate chiral and vector superfields can then be obtained. In order to construct the invariant Lagrangians, we also define chiral densities so that the corresponding action integrals for appropriate chiral and vector superfields remain invariant. Chiral densities are like the square root of the determinant of the metric which one introduces in general theory of relativity for having invariance under general coordinate transformations.

An interesting feature of the supergravity theories is that when one considers such theories, one automatically generates the gravity sector of the model even when starting from chiral superfields which correspond to the "matter" sector. This in a way reminds us of generating gauge invariant interactions while constructing gauge theories. Here as stated in addition to the interaction the "free" gravity term gets generated.

The field operators in the Lagrangian as derived in Chapter 8 are not properly normalised. Further, the description of supergravity here is that of a highly constrained geometric system in superspace, and these constraints are difficult to utilise. In Chapter 9 we give an alternative description of supergravity which is more analogous to the normal formulation of supersymmetry. The Lagrangian obtained here is complicated and is highly nonlinear, but can be applied to obtain phenomenological results more easily. The space of complex scalar fields here form a "Kahler" manifold. For a flat Kahler metric we further show that the supergravity Lagrangian goes over to the conventional supersymmetric Lagrangian when the gravitational coupling vanishes, except for a few terms which have no earlier analogue.

The automatic inclusion of gravity interaction here is the main reason to believe that supergravity in some form may help us to unify all interactions with gravity. However, this feature is also associated with the failure of perturbative renormalisability of the theory. This is in contrast to the attractive feature of supersymmetry, where a cancellation of divergences from the fermionic and bosonic fields occurs. One does not know whether the gain is more or loss is more, and thus the situation is certainly unsatisfactory.

In Chapter 10 we consider some model building with supergravity. Supergravity models have basically two distinct type of approaches. One considers some times supergravity at low energy, i.e. just near the weak scale, and introduces suitable soft symmetry breaking terms.

This relates the phenomenology firmly around the weak scale where the physics is well established. The second approach is through grand unification theories where the soft symmetry breaking terms are derived through spontaneous symmetry breaking at Planck scale. Our discussions here will parallel the second approach above, since it has a better aesthetic appeal.

The most fascinating thing about supergravity is the quantitative link of the phenomena at Planck scale with the same at the weak scale. This scans an energy range of over sixteen orders of magnitude - certainly a magnificent "unification" of physical laws!

Besides emphasizing some of the standard methodology, our analysis of phenomenology will particularly illustrate the interesting features of the symmetry group $SO(10)$. This group absorbs both the left right symmetric semisimple groups of Pati and Salam as well as the minimal $SU(5)$ group of Georgi and Glashow. We also consider a supergravity model based on $SU(5)$ indicating that the Salam Weinberg Higgs particles could be superheavy, e.g. in the grand unification scale. Such an idea is amusing in the context of the present search for the Higgs particles. Further, we briefly note symmetry breaking through radiative corrections with a renormalisation group approach, which puts constraints on the top quark mass. A high top quark mass considered undesirable earlier may as well be a welcome feature in the context of the present experimental situation.

Supergravity theories in higher dimensions are discussed in Chapter 11. We confine our attention to supergravity Lagrangian in ten dimensions. The reason for this is that heterotic string theory in its low energy limit leads to supergravity in ten dimensions. There are also alternative approaches, but here we give only the above to illustrate the ideas of supersymmetry in higher dimensions as well as the compactification of the same to four dimensions. The above as stated is based on superstrings. The nice feature about superstring based theories is that we expect such theories to be renormalisable, a feature we would like to have so that there is a "theory". However, beyond the Planck scale here we have many heavy modes, and the manner in which the overall effect of this can percolate to the four dimensional theory at low energy depends in a nontrivial sense on the way in which the ten dimensional theory is compactified. Some aspects of the same, in the context of string theory or otherwise, is discussed in Chapter 12, along with the remarkable result that Yukawa couplings after compactification can be calculated.

In Chapter 13 we give a very brief description of supersymmetry in two dimensions. This is relevant because much of the model constructions with an idea of understanding the complexities of field theory is usually considered in 1+1 dimensions. We also consider phase transition as an example of vacuum destabilisation where effects of finite temperature are retained. In Chapter 14 we discuss the present status. We also give in appendix A an introduction to Calabi Yau spaces as utilised for the compactification of the ten dimensional space to four observable dimensions, and, in Appendix B, some results in group theory are given.

We note that the theory of supersymmetry and supergravity at present is in a state of flux. E.g. we do not know at a basic level in which dimension it operates, and are ignorant as to whether supergravity is really an input or an output. With our prejudices regarding superstring theories as a "theory of everything", supersymmetry or supergravity could as well be an output instead of being an input. With this in mind we have tried to explain the concepts rather than describe any type of theory in detail which prejudice the beginner.

The quest for unification is the backbone of science. We have therefore thought it desirable to include applications leading to possible unification of all forces. In contrast, there is a school of thought which believes that at present our theoretical as well as experimental knowledge is so limited that the final shape of supersymmetry or supergravity as related to any physical application would be premature. This may as well be true. However, even if it be so, it is necessary to consider some explicit theories. These may later be proved wrong but, will permit us finally to find out what may be right. Instead, trying to extend only mathematical properties may lead to aesthetic beauty without having much relevance for the physical world, just because one is not looking for the same.

We would like to emphasize that a state of flux leads to interesting ideas, some of which will finally emerge to be correct. Our objective here is to give an elementary format with which we may look for this, and understand it as it is discovered. Supersymmetry and supergravity with all its ramifications is after all such a beautiful theory that one feels compelled to believe that nature must use it! We shall go ahead with this "prejudice", since it appears to be the best choice in our quest for unification.

REFERENCES

References [1-6] are papers, [7-9] reviews, [10-14] books.

1. Yu. A. Golfand and E.P. Likhtman, JETP Lett. 13, 323 (1971); D.V. Volkov and V.P. Akulov, Phys. Lett. 46B, 109 (1973).

2. J. Wess and B. Zumino, Nucl. Phys. B70, 39 (1974); ibid B78, 1 (1974).

3. A. Salam and J. Strathdee, Nucl. Phys. B76, 477 (1974).

4. R. Barbieri, S. Ferrara and C. A. Savoy, Phys. Lett. 119B, 343 (1982).

5. A. Chamseddine, R. Arnowitt and Pran Nath, Phys. Rev. Lett. 49, 970 (1982).

6. E. Cremmer, S. Ferrara, L. Girardello and A. Van Proyen, Nucl. Phys. B212, 413 (1983).

7. H.P. Nilles, Phys. Rep. 110, 1 (1984).

8. H. Haber and G. Kane, Phys. Rep. 117, 76 (1984).

9. A.B. Lahanas and D.V. Nanopoulos, Phys. Rep. 145, 1 (1987).

10. J. Wess and J. Bagger, *Introduction to Supersymmetry*, Princeton University Press, Princeton, NJ, 1983.

11. R. Arnowitt, A. Chamseddine and P. Nath, *N=1 Supergravity*, World Scientific, Singapore, 1984.

12. P. West, *Introduction to Supersymmetry and Supergravity*, World Scientific, Singapore, 1986.

13. *Supersymmetry*, edited by S. Ferrara, vol. 1 & 2, World Scientific, Singapore, 1987.

14. *Supersymmetry*, H.J.W. Muller-Kirsten and A. Wiedamann, World Scientific, Singapore, 1987.

Chapter 2

Supersymmetry in Quantum Mechanics

2.1 Grassmann variables

We shall deal with supersymmetry mostly with a superspace formulation, to be defined subsequently. For this purpose, we shall always need Grassmann variables [1], which are really "anticommuting numbers". Thus, if $\theta_1, \theta_2, \cdots, \theta_n$ are n such variables, then

$$\theta_i \theta_j = -\theta_j \theta_i$$

for all i and j. Further, it will be always assumed that these variables anticommute with each other and with any fermionic fields which may be present. They commute with all bosonic variables.

2.2 Superspace in Quantum Mechanics

For the definition of "superspace" in quantum mechanics [2,3] we shall use the triplet $(t, \theta, \bar{\theta})$, where t is the time variable, and θ and $\bar{\theta}$ are two Grassmann variables. We define a super coordinate F as a function of the above three variables. We then in fact obtain the expansion

$$F(t, \theta, \bar{\theta}) = q(t) + \theta \psi(t) + \bar{\psi}(t) \bar{\theta} + \theta \bar{\theta} d(t). \qquad (2.1)$$

We note that the supercoordinate as above is hermitian. In the above equation, the expansion in powers of θ and $\bar{\theta}$ terminates since $\theta^2 = \bar{\theta}^2 = 0$. Thus a supercoordinate in quantum mechanics is expressible in terms of four functions of time, $q(t), \psi(t), \bar{\psi}(t)$ and $d(t)$. We note that a supercoordinate is thus equivalent to four "usual" coordinates given as above. Here, $q(t)$ and $d(t)$ are bosonic, whereas $\psi(t)$ and $\bar{\psi}(t)$ are fermionic.

2.3 Supersymmetry transformation

We take anticommuting parameters $\xi, \bar{\xi}$ such that the supersymmetry transformation δ_ξ in superspace $(t, \theta, \bar{\theta})$ is given as,

$$(t, \theta, \bar{\theta}) \to (t + i\xi\bar{\theta} + i\bar{\xi}\theta, \theta + \xi, \bar{\theta} + \bar{\xi}) \qquad (2.2)$$

Clearly, in the above, t remains hermitian. Here $\xi, \bar{\xi}$ are transalations in superspace and are Grassmann variables [1]. The increment in t as above links ordinary time variable with superspace translations, and our motivation here is to establish such a link. ξ has no dependance on t, so that the above transformation is a global transformation. The generators for the transformation in equation (2.2) as differential operators are,

$$Q_D = \frac{\partial}{\partial \theta} + i\bar{\theta}\frac{\partial}{\partial t} \qquad (2.3a)$$

$$\bar{Q}_D = \frac{\partial}{\partial \bar{\theta}} + i\theta\frac{\partial}{\partial t} \qquad (2.3b)$$

Thus equation (2.2) becomes,

$$\delta_\xi(t, \theta, \bar{\theta}) = (\xi Q_D + \bar{\xi}\bar{Q}_D)(t, \theta, \bar{\theta}). \qquad (2.4)$$

The same also holds for supercoordinate $F(t, \theta, \bar{\theta})$. We then obtain the supersymmetry transformations of the "component" coordinates of equation (2.1) as,

$$\delta_\xi\, q(t) = \xi\psi(t) + \bar{\psi}(t)\bar{\xi}, \qquad (2.5a)$$

$$\delta_\xi\, \psi(t) = \bar{\xi}(-i\dot{q}(t) + d(t)), \qquad (2.5b)$$

$$\delta_\xi\, \bar{\psi}(t) = \xi(d + i\dot{q}(t)), \qquad (2.5c)$$

$$\delta_\xi\, d(t) = i\dot{\bar{\psi}}(t)\bar{\xi} - i\xi\dot{\psi}(t) = \frac{d}{dt}(i\bar{\psi}\bar{\xi} - i\xi\psi). \qquad (2.5d)$$

We thus note that the $\theta\bar{\theta}$ component of a supercoordinate transforms like a total derivative. This is used for the construction of supersymmetric Lagrangian.

2.4 Invariant Lagrangian

Supersymmetry for quantum mechanics was discussed earlier by Witten and by Salomonson and Van Holten [2] where they started with the Lagrangian

$$L = \frac{1}{2}\dot{q}^2 - v(q) + \frac{1}{2}\psi^T(i\dot{\psi} + W(q)\sigma_2\psi) \qquad (2.6)$$

In the above, $\psi = \begin{pmatrix} \psi_1 \\ \psi_2 \end{pmatrix}$. Taking $V(q) = v'(q)$ and $W(q) = v''(q)$ ensures supersymmetry. They also considered quantisation as well as the expression for the supersymmetric charge. We shall approach the same problem through superspace introduced in the last subsection. For this purpose we may define covariant differentiation operators $D, \bar{D}$ given as,

$$D = \frac{\partial}{\partial \theta} - i\bar{\theta}\frac{\partial}{\partial t} \qquad (2.7a)$$

$$\bar{D} = \frac{\partial}{\partial \bar{\theta}} - i\theta\frac{\partial}{\partial t} \qquad (2.7b)$$

which commute with δ_ξ, such that $\delta_\xi D = D\delta_\xi$ and similarly for $\bar{D}$. We shall now write down the Lagrangian density given as, [3]

$$\mathcal{L} = -\frac{1}{2}(\bar{D}F)(DF) - V(F), \qquad (2.8)$$

where the first term denotes the kinetic term of the superspace Lagrangian and second term $V(F)$ is an arbitrary function representing the "superpotential". We now obtain the above Lagrangian density in superspace in terms of the component fields in superspace as

$$\begin{aligned}\mathcal{L} &= (\frac{1}{2}\bar{\psi}\psi - v) + \theta\left[\frac{1}{2}(d+i\dot{q}) - v'\right]\psi + \left[\frac{1}{2}\bar{\psi}(d-i\dot{q}) - v'\right]\bar{\theta} \\ &+ \theta\bar{\theta}\left[\frac{1}{2}\dot{q}^2 + \frac{1}{2}d^2 + \frac{i}{2}(\bar{\psi}\dot{\psi} - \dot{\bar{\psi}}\psi) - dv'\right. \\ &\left. - id\dot{q} + i\dot{q}d - \frac{1}{2}(\psi\bar{\psi} - \bar{\psi}\psi)v''\right]. \end{aligned} \qquad (2.9)$$

One way of dealing with the dynamics above is to consider the coefficient of $\theta\bar{\theta}$ term as in equation (2.8) and then consider the dynamics in the presence of the auxiliary variable d for the quantisation of the above system by eliminating this variable to rewrite the Lagrangian, as in Ref [2]. The quantum mechanical action obtained as the coefficient of $\theta\bar{\theta}$ is 'invariant' under supersymmetry transformations as it always transforms like a total derivative in time. As already stated, we shall however proceed to write the action integral in superspace and then do the quantisation, which brings out some interesting dynamical features in the context of Schwinger's action principle [4] and includes quantisation in addition to getting the equations of motion and the conserved charge. Although we are dealing with quantum mechanics, clearly with this approach equation (2.8) gives the Lagrangian density in *superspace* just as we usually have densities in ordinary space.

2.5 Action principle in Superspace

We shall here illustrate Schwinger's action principle as generalised to superspace very briefly [3,4].

The superspace action integral will necessarily involve an integration in superspace, which includes an integration over the Grassmann variables θ and $\bar{\theta}$. For such integrals we use the two basic formulae given as

$$\int d\theta = 0, \quad \int \theta d\theta = 1. \qquad (2.10)$$

For integrations in fermionic (or Grassmannian) coordinates the above are adequate when we use the conventional formulae of integral calculus, and include the anticommuting properties of the fermionic variables. For simplicity we take one supercoordinate F as above and assume that the superspace Lagrangian density is a function of this supercoordinate as well as of its time derivative and the derivatives with respect to the Grassmannian coordinates. Thus, parallel to Ref.[4] we define the action integral as

$$W_{21} = \int_{t_1}^{t_2} \mathcal{L}\left(F, \frac{\partial F}{\partial t}, \frac{\partial F}{\partial \theta}, \frac{\partial F}{\partial \bar{\theta}}\right) d\theta d\bar{\theta} dt. \qquad (2.11)$$

In the above $\mathcal{L}$ is the Lagrangian density in superspace and the supercoordinate is given as in equation (2.1) in terms of the component coordinates. The equations of motion for the supercoordinates then become

$$\frac{\partial \mathcal{L}}{\partial F} - \frac{\partial}{\partial t}\frac{\partial \mathcal{L}}{\partial \left(\frac{\partial F}{\partial t}\right)} - \frac{\partial}{\partial \theta}\frac{\partial \mathcal{L}}{\partial \left(\frac{\partial F}{\partial \theta}\right)} - \frac{\partial}{\partial \bar{\theta}}\frac{\partial \mathcal{L}}{\partial \left(\frac{\partial F}{\partial \bar{\theta}}\right)} = 0. \qquad (2.12)$$

Such superspace equations combine a number of equations in terms of the component coordinates, and thus have a unified picture for fermionic and bosonic degrees of freedom taken together. Schwinger's

2.5. Action principle in Superspace

action principle is based on changing both the limits of integration as well as giving increments to the fields. Thus, if δt is the change in any limit of integration and δF is the change in the supercoordinate, we then obtain the generators for such changes as [4]

$$G = \int \left[\mathcal{L}\delta t + \delta F \frac{\partial \mathcal{L}}{\partial \left(\frac{\partial F}{\partial t} \right)} \right] d\bar{\theta} d\theta. \qquad (2.13)$$

In deriving the above equation we have used the fact that partial integration of the Grassmannian (fermionic) variables is always possible. The quantisation condition for any supervariable M then becomes [3,4]

$$\left[M, \int \delta F \frac{\delta \mathcal{L}}{\partial \left(\frac{\partial F}{\delta t} \right)} d\bar{\theta} d\theta \right] = i \, \delta M. \qquad (2.14)$$

We shall now illustrate this for the Lagrangian density as in equation (2.8). As in gauge theories, we have constructed the Lagrangian density in superspace in terms of the covariant derivatives there so that supersymmetry is always guaranteed.

2.5.1 Equations of motion

We shall now consider the equations of motion for the Lagrangian density of equation (2.8). In fact in this case we obtain the same in superspace as

$$D \frac{\delta \mathcal{L}}{\delta(DF)} + \bar{D} \frac{\delta \mathcal{L}}{\delta(\bar{D}F)} - \frac{\delta \mathcal{L}}{\delta F} = 0. \qquad (2.15)$$

An explicit evaluation gives the above equation as

$$\frac{1}{2}(D\bar{D} - \bar{D}D)F + v'(F) = 0,$$

which yields that

$$v'(F) + \theta\bar{\theta}F + i\left[\theta \frac{\partial F}{\partial \theta} + \frac{\partial F}{\partial \bar{\theta}}\bar{\theta}\right] + \frac{\partial}{\partial \theta}\frac{\partial}{\partial \bar{\theta}}F = 0. \qquad (2.16)$$

In terms of the component fields we then obtain that

$$d = v'(q), \qquad (2.17a)$$

$$i\dot{\psi} + v''\psi = 0, \quad i\dot{\bar{\psi}} - v''\bar{\psi} = 0, \qquad (2.17b)$$

$$\ddot{q} + d\,v'' + \frac{1}{2}(\psi\bar{\psi} - \bar{\psi}\psi)v''' = 0. \qquad (2.17c)$$

The above equations are obtained by identifying terms with no θ, single θ, single $\bar{\theta}$ and $\theta\bar{\theta}$ respectively. These are exactly the equations obtained by Salomonson and Van Holten in [2], obtained here from the single superspace equation as in (2.16). These equations imply that F of equation (2.1) does not have four degrees of freedom in terms of the component coordinates. We firstly have,

$$\delta d = v''(q)\delta q. \qquad (2.18a)$$

Further, the equations for fermions yields that

$$\frac{d}{dt}(\psi\bar{\psi}) = 0,$$

such that for the corresponding variations we have

$$\delta\psi\,\bar{\psi} + \psi\,\delta\bar{\psi} = 0. \qquad (2.18b)$$

Equations (2.18) demonstrate that variations of coordinates consistent with the equations of motion are really two instead of four. We shall include this fact while doing quantisation. The above conservation rule may be associated with a $U(1)$ symmetry in superspace, R-symmetry, that will be introduced in Chapter 6.

2.5.2 Quantisation

In order to do quantisation, we first note that the "canonically conjugate momentum" in superspace is given as

$$\frac{\partial \mathcal{L}}{\partial \left(\frac{\partial F}{\partial t}\right)} = \frac{i}{2}(\theta\psi - \bar{\psi}\bar{\theta}) + \theta\bar{\theta}\dot{q}. \qquad (2.19)$$

Hence we get the generator G as

$$G = \int \left[(\delta q + \theta\,\delta\psi + \delta\bar{\psi}\,\bar{\theta} + \theta\bar{\theta}\,\delta d)(\frac{i}{2}(\theta\psi - \bar{\psi}\bar{\theta}) + \theta\bar{\theta}\dot{q})\right] d\theta d\bar{\theta}$$

$$= \delta q \dot{q} + \frac{i}{2}(\bar{\psi}\,\delta\psi + \psi\,\delta\bar{\psi}). \qquad (2.20)$$

Hence, writing equation (2.14) in the component form, we obtain

$$[q, \delta q\,\dot{q} + \frac{i}{2}(\bar{\psi}\,\delta\psi + \psi\,\delta\bar{\psi})] = i\delta q, \qquad (2.21a)$$

2.5. Action principle in Superspace

$$[\psi, \delta q\, \dot{q} + \frac{i}{2}(\bar{\psi}\,\delta\psi + \psi\,\delta\bar{\psi})] = i\delta\psi, \qquad (2.21\text{b})$$

$$[\bar{\psi}, \delta q\, \dot{q} + \frac{i}{2}(\bar{\psi}\,\delta\psi + \psi\,\delta\bar{\psi})] = i\delta\bar{\psi}, \qquad (2.21\text{c})$$

$$[d, \delta q\, \dot{q} + \frac{i}{2}(\bar{\psi}\,\delta\psi + \psi\,\delta\bar{\psi})] = i\delta d. \qquad (2.21\text{d})$$

The above equations yield the algebra to be satisfied by the component coordinates, and this has to include the interdependence of the increments as in equations (2.18). This then has the solution

$$[q, \dot{q}] = i, \quad [q, \psi] = 0, \quad [\psi, \bar{\psi}]_+ = 1, \qquad (2.22)$$

and the consistency requirement

$$[d, \dot{q}] = i\delta d, \qquad (2.23)$$

which is also satisfied. We note that the above algebra for the coordinates has been derived from the "canonical" commutation relation as in equation (2.14) for the supercoordinate F when we recognise that the increments concerned are interdependant, and it is nice to note that the superspace algebra combines within itself the bosonic and the fermionic algebra.

2.5.3 Hamiltonian

We already noted that the Lagrangian L is obtained from the Lagrangian density $\mathcal{L}$ in superspace through the integration

$$L = \int \mathcal{L}\, d\bar{\theta} d\theta. \qquad (2.24)$$

We also note that the Hamiltonian density $\mathcal{H}$ in superspace is given as

$$\mathcal{H} = \frac{\partial F}{\partial t} \frac{\partial \mathcal{L}}{\partial\left(\frac{\partial F}{\partial t}\right)} - \mathcal{L}, \qquad (2.25)$$

so that the Hamiltonian H becomes

$$H = \int \mathcal{H}\, d\bar{\theta} d\theta = \frac{1}{2}\left[\dot{q}^2 + v'^2 + (\psi\bar{\psi} - \bar{\psi}\psi)v''\right], \qquad (2.26)$$

where $\dot{q}$ is to be replaced by the canonical conjugate momentum. The above has been used in Ref.[2]. In the above it is worthwhile to note one fact which is peculiar to supersymmetric quantum mechanics. In quantum mechanics, the fermionic level either is empty or is occupied; there is no other degree of freedom for the same. Hence, for the Hamiltonian in equation (2.26), we may replace the fermionic operators by a 2×2 matrix given as [2]

$$V = \frac{1}{2} \begin{bmatrix} v'^2 + v'' & 0 \\ 0 & v'^2 - v'' \end{bmatrix}.$$

The first place in the corresponding vector space implies that there are no fermions, and the second place, that there is a fermion. In case one considers the spectrum of the Hamiltonian, all the levels in the first place will be bosonic, and all the levels in the second place will be fermionic with effectively a potential which is a matrix. This feature has been utilised to obtain many general solutions by Dutt, Khare and Sukhatme [5].

2.5.4 Supersymmetric charges

We first note that as per the basic equation (2.2) a translation in $\theta, \bar{\theta}$ involves a change in time coordinate, and thus in equation (2.11) we have to be more careful regarding the order of integration. With this in mind we rewrite equation (2.11) as

$$W_{21} = \int d\bar{\theta} d\theta \int_{t_1}^{t_2} dt \mathcal{L}. \qquad (2.27)$$

We now take the generators for the supersymmetric transformations, i.e. the supersymmetric charges as $\mathcal{Q}$ and $\bar{\mathcal{Q}}$ defined through

$$\begin{aligned} G_\xi &= \int \left[\mathcal{L}(-i\xi\bar{\theta} - i\bar{\xi}\theta) + \delta_\xi F \frac{\partial \mathcal{L}}{\partial \left(\frac{\partial F}{\partial t}\right)} \right] d\bar{\theta} d\theta \\ &= -i\xi \, \mathcal{Q} - i\bar{\xi} \, \bar{\mathcal{Q}}. \end{aligned} \qquad (2.28)$$

It is to be noted that $\delta\xi = i\xi\bar{\theta} + i\bar{\xi}\theta$ has been neutralised by a $\delta t = -\delta_\xi t$ in equation (2.28) so that after the supersymmetry transformation, time t remains unchanged. This yields that

$$\begin{aligned} \mathcal{Q} &= \int \left[\bar{\theta} \mathcal{L} + i Q_D F \frac{\partial \mathcal{L}}{\partial \left(\frac{\partial F}{\partial t}\right)} \right] d\bar{\theta} d\theta \\ &= (v' + i\dot{q})\psi, \end{aligned} \qquad (2.29)$$

and,

$$\bar{\mathcal{Q}} = \int \left[\theta \mathcal{L} + i \bar{Q}_D F \frac{\partial \mathcal{L}}{\partial \left(\frac{\partial F}{\partial t}\right)}\right] d\bar{\theta} d\theta$$
$$= (v' - i\dot{q})\bar{\psi}, \qquad (2.30)$$

We may note that the supersymmetric charges above flip the zero fermion states to one fermion state and vice versa. We can obtain the conservation of supersymmetric charges from the equation

$$0 = \delta_\xi W_{21}$$
$$= \int d\bar{\theta} d\theta \left[\int_{t_1}^{t_2} dt(\mathcal{L} + \delta\mathcal{L}) - \int_{t_1}^{t_2} dt \mathcal{L}\right]. \qquad (2.31)$$

On using the field equations and some integrations in parts the above equation yields that

$$0 = \delta_\xi W_{21} = \left[-i\xi \mathcal{Q}(t) - i\bar{\xi}\bar{\mathcal{Q}}(t)\right]_{t_1}^{t_2}, \qquad (2.32)$$

where $\mathcal{Q}$ and $\bar{\mathcal{Q}}$ are the supersymmetric charges as defined above. This gives rise to the conservation of the supersymmetric charge. We note that inspite of the supersymmetric transformation, in equation (2.31) t_1 and t_2 remain unaltered since we have taken a compensating δt such that $\delta t + \delta_\xi t = 0$, as in equation (2.28).

We may also note the equation

$$[F, G_\xi] = i\delta_\xi F. \qquad (2.33)$$

The above can be verified from the earlier quantum conditions and also leads to the additonal condition that $\psi^2 = 0$ not derived earlier.

2.6 Remarks

In the above we have attempted to give the picture of supersymmetry in quantum mechanics, introducing the idea of superspace. In particular we have introduced some dynamical features which are most easily understood here. The point to note is that supersymmetry involves auxiliary variables which can be eliminated through the equations of motion, and thus it is a system with constraints. Quantisation of such a system can be done through Dirac bracket [6]. We have however considered it through Schwinger's action principle which aesthetically is a more appealing dynamical framework. We have shown that if we consider the quantisation condition through equation (2.14) and remember that all the increments in the generator G shall not be independent, then quantisation with fermionic variables as well as auxiliary variables becomes possible when a selfconsistent solution of the same can be obtained. It has been seen that in ordinary space also the same thing can be carried out for a supersymmetric Lagrangian [7], along with the fact that from the same Feynman's path integral results can be obtained directly from the Lagrangian without going through the Hamiltonian [7].

Superspace techniques always give a fresh insight to any consideration of supersymmetry. In the next chapters they will be used for the description of supersymmetry in four dimensions. However, the Grassmann variables there not only introduce translations in space-time, but also carry Lorentz transformation properties. This feature introduces a closer link between space-time and the additional fermionic coordinates introduced in the construction of superspace and becomes useful for maintaining Lorentz invariance when we consider supersymmetry.

Superspace is an unphysical construct to obtain a dynamical framework which has a symmetry property for the interchange of fermions and bosons. It would be nice to have a "physical" meaning for this; however, such a meaning can so far only be looked for through the symmetry properties it generates.

REFERENCES

1. F.A. Berezin, *The Theory of Second Quantisation*, Academic Press, New York (1966).

2. P. Salomonson and J.W. Van Holten, Nucl. Phys. B96 509 (1982); E. Witten, Nucl. Phys. B188, 533 (1981).

3. S.P. Misra and T. Pattnaik, Phys. Lett. 129B, 401 (1983).

4. J. Schwinger, Phys. Rev. 82, 914 (1951); 91, 713 (1953).

5. R. Dutt, A. Khare and U.P. Sukhatme, Phys. Lett. 181B, 195 (1986); Am. J. of Phys. 56, 163 (1988).

6. J. Barcelos-Neto and A. Das, Phys. Rev. D33, 2863 (1986).

7. S.P. Misra and T. Pattnaik, Pramana, 24, 595 (1985).

Chapter 3

SL(2,C) Representations of the Lorentz Group

3.1 SL(2,C) and 2-spinors

We shall consider here the Lorentz transformations in the context of the development of supersymmetry. We start with the consideration of the group $SL(2,C)$, which consists of 2×2 complex matrices with determinant as unity. Thus these matrices form a six parameter group, with $\det(a) = 1$ for $a \epsilon SL(2,C)$. We first note that there are four types of two dimensional representations of $SL(2,C)$. These are, (i) $T(a) = a$, (ii) $T(a) = a^*$, (iii) $T(a) = (a^{-1})^{tr}$, and, (iv) $T(a) = (a^{-1})^{tr*}$. Here, a^* denotes complex conjugation. Corresponding to them, we shall use the notations for spinors $\psi_\alpha, \bar{\chi}_{\dot\alpha}, \psi^\alpha$ and $\bar{\chi}^{\dot\alpha}$ as the objects which transform under "a" like

$$\psi'_\alpha = a_{\alpha\beta}\, \psi_\beta \qquad (3.1a)$$

$$\bar{\chi}'_{\dot\alpha} = a^*{}_{\dot\alpha\dot\beta}\, \bar{\chi}_{\dot\beta} \qquad (3.1b)$$

$$\psi'^\alpha = (a^{-1})_{\beta\alpha}\, \psi^\beta \qquad (3.1c)$$

and

$$\bar{\chi}'_{\dot\alpha} = (a^{-1})^*{}_{\dot\beta\dot\alpha}\, \bar{\chi}_{\dot\beta} \qquad (3.1d)$$

Clearly, $\psi^\alpha \psi_\alpha$ and $\bar{\chi}_{\dot\alpha}\bar{\chi}^{\dot\alpha}$ remain invariant under the transformation a. We consistently use $\alpha = 1,2$ for the matrix a and $\dot\alpha = 1,2$ for the complex conjugate matrix a^*.

Let us define the 2×2 matrices σ^m where $\sigma^0 = I$ and σ^i ($i=1,2,3$) are the familiar Pauli spin matrices. We denote their components by

$\sigma^m{}_{\alpha\dot\beta}$. Let us now have a 4-vector P_m ($m=0,1,2,3$) and identify the 2×2 matrix P through

$$P = P_m\,\sigma^m = \begin{bmatrix} P_0 + P_3 & P_1 - iP_2 \\ P_1 + iP_2 & P_0 - P_3 \end{bmatrix}. \qquad (3.2)$$

Clearly $det(P) = P_0{}^2 - P_1{}^2 - P_2{}^2 - P_3{}^2$. Let us transform matrix P under $SL(2,C)$ to get the new matrix

$$P' = a\,P\,a^\dagger. \qquad (3.3)$$

Clearly $detP' = detP$ such that $a\epsilon SL(2,C)$ defines a Lorentz transformation $L\epsilon SO(3,1)$ corresponding to the real transformation

$$P'_m = (L^{-1})_{mn}\,P_n. \qquad (3.4)$$

where P_m are covariant components defined by equation (3.2). We recall that $SL(2,C)$ and $SO(3,1)$ have a two to one correspondence preserving group multiplication. Thus ψ and $\bar\chi_{\dot\alpha}$ are the spinor representations of the Lorentz group. As in Chapters 3 to 7, they constitute the natural objects for the description of supersymmetry transformations in four dimensions. We shall use the antisymmetric tensors $\epsilon^{\alpha\beta}$ and $\epsilon_{\alpha\beta}$ for raising and lowering the indices of ψ_α or ψ^α with the convention that

$$\psi^\alpha = \epsilon^{\alpha\beta}\psi_\beta; \quad \psi_\alpha = \epsilon_{\alpha\beta}\,\psi^\beta. \qquad (3.5)$$

We choose the convention that $\epsilon^{12} = 1 = \epsilon_{21}$. With same definition, $\epsilon^{\dot\alpha\dot\beta}$ and $\epsilon_{\dot\alpha\dot\beta}$ will be used to raise and lower indices of $\bar\chi_{\dot\alpha}$ or $\bar\chi^{\dot\alpha}$. In $SO(3,1)$ we choose the metric as $\eta_{mn} = diag(1,-1,-1,-1)$. We also define the 2×2 matrices $\bar\sigma^m = (I, -\vec\sigma)$ with the components of such matrices written as $\bar\sigma^{m\dot\alpha\beta}$. We use the notations that

$$\psi\chi = \psi^\alpha\chi_\alpha, \qquad (3.6a)$$

$$\bar\chi\bar\psi = \bar\chi_{\dot\alpha}\,\bar\psi^{\dot\alpha}, \qquad (3.6b)$$

$$\psi\sigma^m\bar\chi = \psi^\alpha\sigma^m{}_{\alpha\dot\beta}\,\bar\chi^{\dot\beta}, \qquad (3.6c)$$

and

$$\bar\chi\bar\sigma^m\psi = \bar\chi_{\dot\alpha}\,\bar\sigma^{m\dot\alpha\beta}\,\psi_\beta. \qquad (3.6d)$$

In what follows, ψ^α, ψ_α or $\bar\chi_{\dot\alpha}$, $\bar\chi^{\dot\alpha}$ denote Weyl spinors as functions of space time variables x^m. In addition, we shall also take the spinor doublets θ^α or $\bar\theta^{\dot\alpha}$ as merely Grassmannian variables, which are also Weyl spinors.

We may now easily note the following identities:

3.1. SL(2,C) and 2-spinors

$$\theta^\alpha \theta^\beta = -\frac{1}{2} \epsilon^{\alpha\beta} (\theta\theta), \tag{3.7a}$$

$$\theta_\alpha \theta_\beta = \frac{1}{2} \epsilon_{\alpha\beta} (\theta\theta), \tag{3.7b}$$

$$\bar{\theta}^{\dot\alpha} \bar{\theta}^{\dot\beta} = \frac{1}{2} \epsilon^{\dot\alpha\dot\beta} (\bar\theta\bar\theta), \tag{3.7c}$$

$$(\theta\sigma^m\bar\theta)(\theta\sigma^n\bar\theta) = \frac{1}{2} \eta^{mn} (\theta\theta)(\bar\theta\bar\theta), \tag{3.7d}$$

$$\bar\sigma^{m\dot\alpha\beta} = \epsilon^{\dot\alpha\dot\gamma} \epsilon^{\beta\gamma} \sigma^m{}_{\gamma\dot\gamma}, \tag{3.7e}$$

$$\eta_{mn} \sigma^m{}_{\alpha\dot\alpha} \sigma^n{}_{\beta\dot\beta} = 2 \epsilon_{\alpha\beta} \epsilon_{\dot\alpha\dot\beta}, \tag{3.7f}$$

$$Tr(\sigma^m \bar\sigma^n) = 2 \eta^{mn}, \tag{3.7g}$$

$$(\psi\chi) = (\chi\psi), \tag{3.7h}$$

$$\psi\sigma^m\bar\chi = -\bar\chi\bar\sigma^m\psi, \tag{3.7i}$$

$$(\psi\sigma^m\bar\chi)^\dagger = \chi\sigma^m\bar\psi, \tag{3.7j}$$

$$(\sigma^m\bar\sigma^n + \sigma^n\bar\sigma^m)_\alpha{}^\beta = 2\eta^{mn}\delta_\alpha{}^\beta, \tag{3.7k}$$

$$(\bar\sigma^m\sigma^n + \bar\sigma^n\sigma^m)^{\dot\alpha}{}_{\dot\beta} = 2\eta^{mn}\delta^{\dot\alpha}{}_{\dot\beta}. \tag{3.7l}$$

In equations (3.7h) and (3.7i) we have assumed that the respective spinorial fields anticommute.

3.2 Dirac Spinors

We now define the four component spinors as follows:

$$\psi_\alpha^{(D)}\big|_{\text{four components}} = \begin{pmatrix} \phi_\alpha \\ \bar\chi^{\dot\alpha} \end{pmatrix}. \qquad (3.8)$$

On the left hand side of equation (3.8) in $\psi_\alpha^{(D)}$, α runs over four components, whereas, for Weyl spinors ϕ_α and $\bar\chi^{\dot\alpha}$, α runs over two components. In the above, $\psi_\alpha^{(D)}$ is a Dirac spinor, which consists of a pair of Weyl spinors, and thus belongs to the representations $D(1/2,0)$ and $D(0,1/2)$ respectively for the two component spinors ϕ_α and $\bar\chi^{\dot\alpha}$. (We use the same α to indicate both two as well as four components. This will cause no confusion as things will be clear from the context.) Let us now introduce 4×4 Dirac matrices given as:

$$\gamma^m = \begin{bmatrix} 0 & \sigma^m \\ \bar\sigma^m & 0 \end{bmatrix}. \qquad (3.9)$$

We then have, from equations (3.7j) and (3.7k),

$$\gamma^m \gamma^n + \gamma^n \gamma^m = 2\eta^{mn}, \qquad (3.10)$$

where there is a 4×4 identity matrix implicit on the right hand side of the equation (3.10).

Let us define

$$\gamma_5 = \begin{pmatrix} I & 0 \\ 0 & -I \end{pmatrix}. \qquad (3.11)$$

We then easily see that the spinors

$$\psi_\pm^{(W)} = \frac{1}{2}(I \pm \gamma_5)\psi^{(D)} \qquad (3.12)$$

are trivially four component Weyl spinors as chiral projections of the Dirac spinor to ϕ_α and $\bar\chi^{\dot\alpha}$ respectively.

The four component spinor

$$\psi^{(M)} = \frac{1}{\sqrt{2}} \begin{pmatrix} \phi_\alpha \\ \bar\phi^{\dot\alpha} \end{pmatrix} \qquad (3.13)$$

is known as a Majorana spinor. Here the field for the representation $D(0,1/2)$ is not an independent field, and thus associated with Dirac equation we shall not have two degrees of freedom for particles and antiparticles. The Grassmannian variables θ's as four component objects are in fact

3.2. Dirac Spinors

$$\theta^{(M)} = \begin{pmatrix} \theta_\alpha \\ \bar{\theta}^{\dot{\alpha}} \end{pmatrix} \qquad (3.14)$$

and thus are Majorana spinors.

We may also have raising and lowering matrices for 4-spinors. Thus, we define the 4×4 matrix

$$C^{(D)\alpha\beta} = \begin{pmatrix} \epsilon^{\alpha\beta} & 0 \\ 0 & \epsilon_{\dot{\alpha}\dot{\beta}} \end{pmatrix} \qquad (3.15a)$$

and

$$C^{(D)}_{\alpha\beta} = \begin{pmatrix} \epsilon_{\alpha\beta} & 0 \\ 0 & \epsilon^{\dot{\alpha}\dot{\beta}} \end{pmatrix} \qquad (3.15b)$$

We may then write that

$$\psi^{(D)\alpha} = C^{(D)\alpha\beta} \psi_\beta. \qquad (3.16)$$

We may verify that $\psi^{(D)\alpha}\psi^{(D)}_\alpha$ is Lorentz invariant. We also note that α, β on the left hand side of equations (15) take four values, whereas the same indices on right hand side of the equation take two values, as would always be clear from the context.

The adjoint of the Dirac operator is given as

$$\bar{\psi}^{(D)tr} = \left(\psi^{(D)\,\dagger}\gamma^0\right)^{tr} = \begin{pmatrix} \chi^\alpha \\ \bar{\phi}_{\dot{\alpha}} \end{pmatrix}. \qquad (3.17)$$

The relationship

$$v_{\alpha\dot{\alpha}} = v_m \sigma^m{}_{\alpha\dot{\alpha}}, \qquad (3.18a)$$

and its inverse

$$v^m = \frac{1}{2}\bar{\sigma}^{m\dot{\alpha}\alpha} v_{\alpha\dot{\alpha}} \qquad (3.18b)$$

will permit us to go freely from tensors to a product representations in spinor space and vice versa. Also with reference to equation (3.8), we may note the charge conjugation operation as

$$^{(c)}\psi^{(D)} = -i\gamma^2 \psi^{(D)*} = \begin{pmatrix} \chi_\alpha \\ \bar{\phi}^{\dot{\alpha}} \end{pmatrix}. \qquad (3.19)$$

The relevance of this in the context of equation (3.13) is obvious, since clearly $^{(c)}\psi^{(M)} = \psi^{(M)}$ i.e. the Majorana field is self charge-conjugate.

We may have many alterntive forms for the matrices satisfying the algebra (3.10), which are related to each other by similarity transformations, as is well known. Here we have merely thought it useful to explain the above notations which make the two component and four component treatments interchangeable with ease, along with a brief description of the ideas for the sake of completeness.

REFERENCES

1. J. Wess and J. Bagger, *Introduction to Supersymmetry*, Princeton University Press, NJ, 1983.

Chapter 4

Superfields

4.1 Introduction

Supersymmetry is a beautiful and systematic concept with integrated bosonic and fermionic symmetries. A nice way of defining field theory here is to proceed through superfields as was proposed by Salam and Strathdee [1]. We first define superspace with coordinates defined as $(x^m, \theta^\alpha, \bar{\theta}_{\dot{\alpha}})$ where x^m are ordinary spacetime variables and, θ and $\bar{\theta}$ are two component Grassmann variables which transform as Weyl spinors. Superfields shall be functions defined on superspace. Let us now define a transformation in superspace given as

$$(x^m, \theta, \bar{\theta}) \to (x^m + i\theta\sigma^m\bar{\xi} - i\xi\sigma^m\bar{\theta},\ \theta + \xi,\ \bar{\theta} + \bar{\xi}) \tag{4.1}$$

where ξ and $\bar{\xi}$ are anticommuting increments describing translation in superspace for the Grassmannian coordinates, and, simultaneously yielding a translation in space-time. We may compare this equation with equation (2.2) in quantum mechanics, where we had only time coordinate t. In equation (4.1) clearly ξ and $\bar{\xi}$ are Weyl spinors, and as earlier they do not depend on space-time which is equivalent to saying that the transformations are global. Parallel to equation (2.2), the transformations here mix up Grassmann coordinates with space-time coordinates, which will help us to define supersymmetry transformations for the field operators.

We can write this transformation in terms of generators as

$$\delta_\xi = \xi^\alpha Q_\alpha + \bar{\xi}_{\dot{\alpha}} \bar{Q}^{\dot{\alpha}} \tag{4.2}$$

with

$$Q_\alpha = \frac{\partial}{\partial \theta^\alpha} - i\sigma^m{}_{\alpha\dot{\alpha}}\bar{\theta}^{\dot{\alpha}}\, \partial_m\,, \tag{4.3a}$$

$$\bar{Q}_{\dot\alpha} = -\frac{\partial}{\partial \bar\theta^{\dot\alpha}} + i\theta^\alpha \sigma^m{}_{\alpha\dot\alpha}\, \partial_m\,. \tag{4.3b}$$

which are parallel of equations (2.3). We further have the algebra

$$Q_\alpha \bar Q_{\dot\alpha} + \bar Q_{\dot\alpha} Q_\alpha = 2\sigma^m{}_{\alpha\dot\alpha} P_m, \tag{4.4}$$

where $P_m = i\,\partial_m = i\frac{\partial}{\partial x^m}$ denotes the covariant momentum components in space. We note that raising and lowering of the indices is done here through $\epsilon^{\alpha\beta}$ and $\epsilon_{\alpha\beta}$ matrices. We shall now define a superfield $F(x,\theta,\bar\theta)$ [2] which can be expanded in terms of θ and $\bar\theta$ given as,

$$\begin{aligned}
F(x,\theta,\bar\theta) &= f(x) + \theta\psi(x) + \bar\theta\bar\phi(x) + (\theta\theta)\,m(x) + (\bar\theta\bar\theta)n(x) \\
&\quad + \theta\sigma^m\bar\theta v_m(x) + (\theta\theta)\,\bar\theta\bar\lambda(x) + (\bar\theta\bar\theta)\,\theta\chi(x) \\
&\quad + (\theta\theta)(\bar\theta\bar\theta)\,d(x)
\end{aligned} \tag{4.5}$$

where the notations are as per the last chapter. We have chosen the superfield operator above as a Lorentz scalar. The fields $f, \psi_\alpha, \bar\phi^{\dot\alpha}, m, n, v_m, \bar\lambda^{\dot\alpha}, \chi_\alpha$, and d are called component fields of the superfield, and have obvious Lorentz transformation properties. Here the series terminates after a finite number of terms due to the anticommuting nature of θ's. We may note that there are sixteen bosonic and fermionic degrees of freedom since all the fields under consideration are here complex. We now define the supersymmetry transformation for the superfield $F(x,\theta,\bar\theta)$ by equation (4.2) as,

$$\begin{aligned}
\delta_\xi F(x,\theta,\bar\theta) &= F(x + \delta_\xi x, \theta + \delta_\xi \theta, \bar\theta + \delta_\xi \bar\theta) - F(x,\theta,\bar\theta) \\
&\equiv \delta_\xi f(x) + \theta^\alpha\, \delta_\xi \psi_\alpha(x) + \cdots,
\end{aligned} \tag{4.6}$$

where all the other terms on the right hand side of equation (4.5) are to be continued in (4.6). We can now calculate the transformation laws for the component fields, some of which are quoted below:

$$\delta_\xi f(x) = \xi^\alpha \psi_\alpha(x) + \bar\xi_{\dot\alpha} \bar\phi^{\dot\alpha}(x), \tag{4.7a}$$

$$\delta_\xi \psi_\alpha(x) = i\sigma^m{}_{\alpha\dot\alpha}\,\bar\xi^{\dot\alpha}\, \partial_m f(x) + 2\,\xi_\alpha\, m(x), \tag{4.7b}$$

$$\delta_\xi \bar\phi^{\dot\alpha}(x) = i\bar\sigma^{m\dot\alpha\alpha}\, \xi_\alpha\, \partial_m f(x) + 2\,\bar\xi^{\dot\alpha}\, n(x), \tag{4.7c}$$

$$\delta_\xi d(x) = -\frac{i}{2}\, \partial_m \left[\chi(x)\sigma^m\bar\xi + \bar\lambda(x)\bar\sigma^m\xi\right]. \tag{4.7d}$$

These component fields will transform such that they map boson fields into fermion fields and vice versa. However, the field $d(x)$ transforms like a four divergence, a property which will be used to construct the supersymmetric Lagrangian. We may note that such superfields $F(x,\theta,\bar\theta)$ are reducible, i.e. extra component fields can be eliminated through covariant constraints. We shall e.g. have two types of superfields depending on specific constraints, i.e. the chiral and vector superfields, which we shall discuss in the next sections.

4.2 Chiral Superfields

Superfields Φ and $\Phi^\dagger$ satisfying the condition $\bar{D}\Phi = 0$ and $D\Phi^\dagger = 0$, are called chiral and antichiral superfields respectively, where $D, \bar{D}$ are operators in superspace given as

$$D_\alpha = \frac{\partial}{\partial \theta^\alpha} + i\sigma^m{}_{\alpha\dot{\alpha}}\bar{\theta}^{\dot{\alpha}}\partial_m, \qquad (4.8a)$$

$$\bar{D}_{\dot{\alpha}} = -\frac{\partial}{\partial \bar{\theta}^{\dot{\alpha}}} - i\theta^\alpha \sigma^m{}_{\alpha\dot{\alpha}}\partial_m, \qquad (4.8b)$$

These differential operators anticommute with the supersymmetric generators Q_α and $\bar{Q}_\alpha$, and hence are called covariant differentiation operators with respect to this symmetry. Further, the condition (4.8) implies that the superfields Φ, $\Phi^\dagger$ describe irreducible multiplets of ordinary fields. We can show this as follows. Let us first note that if we substitute

$$y^m = x^m + i\theta\sigma^m\bar{\theta}, \qquad (4.9)$$

from equation (4.8b) we clearly have $\bar{D}_{\dot{\alpha}}y^m = 0$. Further, it is obvious that we also have $\bar{D}_{\dot{\alpha}}\theta^\beta = 0$. Hence when we take the superfield $\Phi = \Phi(y,\theta)$ as a function of y and θ, we shall have $\bar{D}_{\dot{\alpha}}\Phi(y,\theta) = 0$ and thus such a superfield Φ is a chiral superfield. We may now write

$$\Phi(y,\theta) = A(y) + \sqrt{2}\,\theta^\alpha \psi_\alpha(y) + (\theta\theta)F(y), \qquad (4.10)$$

where $A(y)$ is a complex scalar field, $\psi_\alpha(y)$ is a spinor field and $F(y)$ is the auxiliary field.

In terms of variables y and θ, we have in fact

$$Q_\alpha = \frac{\partial}{\partial \theta^\alpha}, \qquad (4.11a)$$

$$\bar{Q}_{\dot{\alpha}} = -\frac{\partial}{\partial \bar{\theta}^{\dot{\alpha}}} + 2i\theta^\alpha \sigma^m{}_{\alpha\dot{\alpha}}\frac{\partial}{\partial y^m}. \qquad (4.11b)$$

Then one can clearly see that the component fields $A(y)$, $\psi(y)$ and $F(y)$ have the corresponding supersymmetric transformations given as

$$\delta_\xi A(y) = \sqrt{2}\,\xi\psi(y), \qquad (4.12a)$$

$$\delta_\xi \psi_\alpha(y) = \sqrt{2}\,i\sigma^m{}_{\alpha\dot{\alpha}}\bar{\xi}^{\dot{\alpha}}\partial_m A + \sqrt{2}\,\xi_\alpha F, \qquad (4.12b)$$

$$\delta_\xi F(y) = \sqrt{2}\,i\bar{\xi}\bar{\sigma}^m \partial_m \psi. \qquad (4.12c)$$

These transformation laws agree with (4.7) as obtained for a general superfield and show that A, ψ and F are closed with respect to supersymmetry transformations. We can also express the chiral superfield Φ in terms of the supercoordinates $z \equiv (x, \theta, \bar{\theta})$ to get

$$\begin{aligned}\Phi(x,\theta,\bar{\theta}) &= A(x) + i\theta\sigma^m\bar{\theta}\, \partial_m\, A(x) - \frac{1}{4}(\theta\theta)(\bar{\theta}\bar{\theta})\Box A(x) \\ &+ \sqrt{2}\,\theta\psi(x) - \frac{i}{\sqrt{2}}(\theta\theta)\,\partial_m\,\psi(x)\sigma^m\bar{\theta} + (\theta\theta)F(x).\end{aligned}$$

(4.13)

4.3 Vector Superfields

The superfields satisfying the reality condition $V(x,\theta,\bar{\theta})^\dagger = V(x,\theta,\bar{\theta})$ are called the vector superfields. In the usual manner one can obtain the components of $V(x,\theta,\bar{\theta})$ given as

$$\begin{aligned}V(x,\theta,\bar{\theta}) &= C(x) + i\theta\chi(x) - i\bar{\theta}\bar{\chi}(x) + \frac{i}{2}(\theta\theta)(M(x)+iN(x)) \\ &- \frac{i}{2}(\bar{\theta}\bar{\theta})(M(x)-iN(x)) - \theta\sigma^m\bar{\theta}v_m(x) + i(\theta\theta)\,\bar{\theta}\bigl[\bar{\lambda}(x) \\ &+ \frac{i}{2}\bar{\sigma}^m\,\partial_m\,\chi(x)\bigr] - i(\bar{\theta}\bar{\theta})\,\theta\bigl[\lambda(x) + \frac{i}{2}\sigma^m\,\partial_m\,\bar{\chi}(x)\bigr] \\ &+ \frac{1}{2}(\theta\theta)(\bar{\theta}\bar{\theta})(D(x)+\frac{1}{2}\Box C(x)).\end{aligned}$$

(4.14)

Here the bosonic component fields C, D, M, N, and $v_m(x)$ are all real. We may note here that the supersymmetric transformation of the component fields is rather complicated and the number of field is rather large. One can simplify it by reducing the number of such components through some "covariant" constraint. For this purpose let us investigate the transformation of each field under a transformation defined in terms of superfields as,

$$V(z) \to V(z) + \Lambda(z) + \Lambda^\dagger(z),$$

(4.15)

where, as stated earlier $z = (x, \theta, \bar{\theta})$ stands for the coordinates in superspace, $\Lambda(z)$ is a classical chiral field and, $\Lambda(z)^\dagger$ is the corresponding antichiral field. Thus $\Lambda(z) + \Lambda(z)^\dagger$ is a vector superfield and the above transformation is in the space of vector superfields. We shall see that this transformation will define a supersymmetric generalisation of abelian gauge transformations. With the expansion of $\Lambda(z)$ as per equation (4.13), the superfield expansion of $\Lambda(z) + \Lambda(z)^\dagger$ is given as

4.3. Vector Superfields

$$\begin{aligned}\Lambda(z) + \Lambda^\dagger(z) &= A(x) + A(x)^* + \sqrt{2}\,(\theta\psi(x) + \bar{\theta}\bar{\psi}(x)) + (\theta\theta)F(x) \\ &+ (\bar{\theta}\bar{\theta})F(x)^* + i\theta\sigma^m\bar{\theta}\,\partial_m\,(A(x) - A(x)^*) \\ &+ \frac{i}{\sqrt{2}}\,(\theta\theta)\,\bar{\theta}\bar{\sigma}^m\,\partial_m\,\psi(x) + \frac{i}{\sqrt{2}}(\bar{\theta}\bar{\theta})\theta\sigma^m\,\partial_m\,\bar{\psi}(x) \\ &- \frac{1}{4}\,(\theta\theta)\,(\bar{\theta}\bar{\theta})\Box(A(x) + A(x)^*).\end{aligned} \qquad (4.16)$$

Hence the gauge transformations of the component fields of $V(x,\theta,\bar{\theta})$ are given as

$$C(x) \to C(x) + A(x) + A(x)^*, \qquad (4.17a)$$

$$\chi(x) \to \chi(x) - i\sqrt{2}\,\psi(x), \qquad (4.17b)$$

$$M(x) + iN(x) \to M(x) + iN(x) - 2iF(x), \qquad (4.17c)$$

$$v_m(x) \to v_m(x) - i\,\partial_m\,(A(x) - A(x)^*), \qquad (4.17d)$$

$$\lambda(x) \to \lambda(x), \qquad (4.17e)$$

and,

$$D(x) \to D(x). \qquad (4.17f)$$

Thus the fields λ and D remain invariant under the gauge transformation as defined above. Also equation (4.15) implies that with a "special" gauge transformation

$$C(x) = -(A(x) + A(x)^*), \qquad (4.18a)$$

$$\chi(x) = i\sqrt{2}\,\psi(x), \qquad (4.18b)$$

$$M(x) + iN(x) = 2iF(x), \qquad (4.18c)$$

the above vector superfield as a multiplet of component fields reduces to the form, $V = (0,0,0,0,v_m,\lambda,D)$. This special gauge is known as the Wess-Zumino gauge. We now call the component fields v_m as the gauge field, λ as the gaugino field, and, $D(x)$ as the auxiliary field as we shall see later that this is not a dynamical field and can be eliminated with the equations of motion. This terminology is in the context of local gauge symmetry which will be considered in the next chapter. It may be noted here that this "gauge fixation" apparently breaks supersymmetry. We can verify this fact by calculating the supersymmetry transformation through

$$\delta_\xi V = (\xi^\alpha Q_\alpha + \bar{\xi}_{\dot\alpha} \bar{Q}^{\dot\alpha})V$$

using equations (4.3) which will generate terms of V inconsistent with the choice of Wess-Zumino gauge. We can rectify this situation by following up every supersymmetry transformation with a suitable gauge transformation to restore Wess-Zumino gauge, as in equations (4.18). E.g. we may take the field dependant compensating gauge transformation in terms of the supersymmetry parameters as

$$\Lambda(z) = i\sqrt{2}\,\theta\sigma^m\bar{\xi}v_m - i/2(\theta\theta)\bar{\lambda}\bar{\xi} \qquad (4.19)$$

which restores the Wess-Zumino gauge after the supersymmetric transformation. This product transformation is always available for a Lagrangian which is both gauge invariant and supersymmetric invariant and is utilised to define a fresh supersymmetry transformation. We also note that when the vector superfield is chosen to be in Wess-Zumino gauge, the component fields C, χ, M, N are zero and under a "gauge transformation" given by $\Lambda(z)$ we would naturally like that Wess-Zumino gauge may be maintained. For this to happen, the chiral gauge superfield $\Lambda(z)$ can not be arbitrary. In fact, from equations (4.17) we realise that here the complex scalar field $A(x)$ shall be pure imaginary, and the fields $\psi(x)$ and F(x) must be zero. Hence in order that Wess-Zumino gauge be maintained, the permissible gauge transformations get fully defined through a *single* real function of space-time corresponding to $U(1)$ gauge transformations as conventionally taken. We note that for this group as in (4.17) the vector field transforms in the conventional manner, where as the gaugino field *does not* transform. The above constraint for the chiral field which defines the gauge transformation, as well as the interplay of gauge transformation with supersymmetry as utilised to define the new set of supersymmetry transformations, should be noted as special features of Wess-Zumino gauge which permits the number of component fields to be minimal and still have a well-defined gauge symmetry and supersymmetry.

4.4 Supersymmetric Lagrangian

4.4.1 Chiral Superfields

In this section we shall first construct the Lagrangian for chiral superfields. Usually it consists of two types of terms, the F-type and the D-type of terms, as shall be subsequently constructed out of the product of superfields. In general it contains a kinetic term, a mass term, Yukawa interactions of fermions with scalar fields and, cubic and quartic couplings of scalar fields. They may be generated as per the following prescriptions. Let us take

4.4. Supersymmetric Lagrangian

$$\mathcal{L} = \mathcal{L}_{\text{kin}} + \mathcal{L}_{\text{mass}} + \mathcal{L}_{\text{int}} , \qquad (4.20)$$

where,

$$\mathcal{L}_{\text{kin}} = \Phi(x,\theta,\bar{\theta})\Phi^{\dagger}(x,\theta,\bar{\theta})\Big|_{(\theta\theta)(\bar{\theta}\bar{\theta})} , \qquad (4.21a)$$

$$\mathcal{L}_{\text{mass}} = \frac{m}{2}\left(\Phi(y,\theta)\right)^2\Big|_{(\theta\theta)y\to x} + h.c. \qquad (4.21b)$$

$$\mathcal{L}_{\text{int}} = \frac{1}{3!}\, g\left(\Phi(y,\theta)\right)^3\Big|_{(\theta\theta)y\to x} + h.c. \qquad (4.21c)$$

In the above, m is the mass and g, the coupling constant. Further, $(\theta\theta)(\bar{\theta}\bar{\theta})$ terms are known as D-type terms, whereas $(\theta\theta)$ or $(\bar{\theta}\bar{\theta})$ terms are F-type terms. Thus, (4.21a) is D-type, whereas the terms (4.21b) and (4.21c) are F-type.

A few remarks regarding the mass dimensions of the terms in equations (4.21) may be worthwhile. We note that from equation (4.5) or (4.10), the concerned chiral superfields have mass diemension one, i.e. the same as that of scalar fields. Further, the mass dimensions of the Grassmannian supercoordinates θ and $\bar{\theta}$ are $-1/2$ as is obtained from the canonical mass dimension of $3/2$ for the fermion fields. This gives that the mass dimensions of the auxiliary fields is two. Hence the mass dimensions of the expression on the right hand side of equation (4.21a) is now seen to be $2 + 4 \times (1/2) = 4$ as is expected. Further, clearly the parameter m in equation (4.21b) shall have the dimension of mass, and, the parameter g in equation (4.21c) shall be dimensionless, so that the dimensions of the Lagrangian densities in the above equations remain as expected. From the equations of motion we shall later show that m is the mass of the supersymmetric particles, and that g is the Yukawa coupling constant.

A few other remarks regarding equations (4.21) are also relevant. We firstly note that products of superfields are superfields. Further, as stated in the last one of the equations (4.7), the coefficient of $(\theta\theta)(\bar{\theta}\bar{\theta})$ for any superfield transforms under a supersymmetry transformation as a four divergence. Hence the contribution in (4.21a), or for that matter any D-type contribution, shall be invariant under a supersymmetry transformation except for a four divergence. Further, we note that any product of chiral superfield is a chiral superfield, and that the coefficient of the $(\theta\theta)$ term of a chiral superfield again always transforms as a four divergence as given in the last of the equations (4.12). Hence the Lagrangian contributions of equations (4.21b) and (4.21c), or for that matter any F-type contribution, shall be invariant under supersymmetry transformations again except for a four divergence. Hence the total Lagrangian constructed in equations (4.21) is invariant under supersymmetry transformations, modulo four divergence. Such a description of invariance is a new feature of supersymmetry. For a chiral multiplet thus we have constructed a supersymmetric Lagrangian; we shall now show that the Lagrangian as constructed is dynamically relevant. We shall now substitute the component field expression (4.13) for the chiral superfields in equations (4.21). Then the Lagrangian in terms of component fields is given as,

$$\mathcal{L}_{\text{kin}} = \left[-\frac{1}{4} A^* \Box A - \frac{1}{4} \Box A^* A + \frac{1}{2} \partial_m A^* \partial^m A \right.$$
$$\left. + \frac{i}{2} (\partial_m \bar{\psi} \bar{\sigma}^m \psi - \bar{\psi} \bar{\sigma}^m \partial_m \psi) \right] + F^* F$$
$$\equiv (\partial_m A^*)(\partial^m A) + i(\bar{\psi} \bar{\sigma}^m \partial_m \psi) + F^* F, \qquad (4.22a)$$

$$\mathcal{L}_{\text{mass}} = m[AF + A^* F^*] - \frac{m}{2}(\psi\psi + \bar{\psi}\bar{\psi}) \qquad (4.22b)$$

$$\mathcal{L}_{\text{int}} = g\,[A^2 F + A^{*\,2} F^* - (\psi\psi)A - (\bar{\psi}\bar{\psi})A^*]. \qquad (4.22c)$$

Here as mentioned earlier, F and F^* are auxiliary fields which are given through the equation

$$\frac{\partial \mathcal{L}}{\partial F^*} = F + mA^* + gA^{*\,2} = 0. \qquad (4.23)$$

We now substitute $F = -(mA^* + gA^{*2})$ in the Lagrangian expression to obtain

$$\mathcal{L} = (\partial_m A)^*(\partial^m A) - m^2 A^* A + i\,\partial_m \bar{\psi}\bar{\sigma}^m \psi - \frac{m}{2}[\psi\psi + \bar{\psi}\bar{\psi}]$$
$$- g\,(\psi\psi A + \bar{\psi}\bar{\psi}A^*) - mg\,A^* A(A^* + A) - g^2 A^{*2} A^2. \qquad (4.24)$$

In deriving equation (4.24) we have permitted ourselves to do a few integrations in parts to write the Lagrangian in a convenient form. The dynamical relevance of the above Lagrangian is obvious. We easily identify the terms on the right hand side above as the kinetic term for the complex scalar field, its mass term, the kinetic term of the spin half field and its mass term, Yukawa coupling of the spin half field, and, the cubic and quartic couplings of the complex scalar field with itself. All the interaction terms are described in terms of a single coupling constant, so that the dynamical contributions from bosons and fermions get related (leading some times to the cancellation of divergences), and, the contributions for external boson or fermion lines also get related, leading to supersymmetry for explicit dynamical reactions.

It shall be instructive to analyse the Yukawa coupling term

4.4. Supersymmetric Lagrangian

$$\mathcal{L}_{\text{Yukawa}} = -g(\,\psi\psi\,A + \bar{\psi}\bar{\psi}\,A^*) \tag{4.25}$$

in the context of the real and imaginary parts of the complex scalar field A. Let us consider the Majorana field operator

$$\psi^{(M)} = \frac{1}{\sqrt{2}}\begin{pmatrix} \psi_\alpha(x) \\ \bar{\psi}^{\dot\alpha}(x) \end{pmatrix}. \tag{4.26}$$

With equations (4.11) and (4.16), we then obtain that

$$\bar{\psi}^{(M)}\psi^{(M)} = \frac{1}{2}(\psi\psi + \bar{\psi}\bar{\psi}) \tag{4.27a}$$

and

$$\bar{\psi}^{(M)}\gamma^5\psi^{(M)} = \frac{1}{2}(\psi\psi - \bar{\psi}\bar{\psi}). \tag{4.27b}$$

Hence clearly (4.25) yields that

$$\mathcal{L}_{\text{Yukawa}} = -g\sqrt{2}\,(\bar{\psi}^{(M)}\psi^{(M)}\,S + \bar{\psi}^{(M)}\gamma^5\psi^{(M)}\,P), \tag{4.28}$$

where

$$S = \frac{1}{\sqrt{2}}(A + A^*), \quad P = \frac{1}{\sqrt{2}}(A - A^*). \tag{4.29}$$

Thus, clearly, the real part of A is a scalar field whereas the imaginary part of A is a pseudoscalar field.

We may generalise the above concept with the introduction of a function of chiral superfields (which is also necessarily a chiral superfield) and collect $(\theta\theta)$ term from it. Thus we may replace equation (4.21c) by the equation

$$\mathcal{L}_{\text{int}} = g(\Phi)|_{(\theta\theta)} + h.c. \tag{4.30}$$

where, $g(\Phi)$ is a function of a class of chiral superfields Φ_i. This yields the most general F-type term of the Lagrangian for the chiral superfields. We shall recognise later that renormalisability of supersymmetric Lagrangian demands that $g(\Phi)$ may be at most a cubic expression in Φ. $g(\Phi)$ is called the *superpotential*. With the arbitrary number of chiral superfields Φ, we thus have,

$$\begin{aligned}
\mathcal{L}\big|_{\text{F-type}} &= g(\Phi)\big|_{(\theta\theta)} + h.c. \\
&= \left[g(A) + \frac{\partial g}{\partial A_i}(\sqrt{2}\,\theta\psi_i + (\theta\theta)F_i) \right. \\
&\quad \left. + \frac{1}{2}\frac{\partial^2 g}{\partial A_i \partial A_j}(\sqrt{2}\,\theta\psi_i + (\theta\theta)F_i)(\sqrt{2}\,\theta\psi_j + (\theta\theta)F_j) \right]_{\theta\theta} + h.c. \\
&= \frac{\partial g}{\partial A_i}F_i - \frac{1}{2}\frac{\partial^2 g}{\partial A_i \partial A_j}(\psi_i\psi_j) + h.c.
\end{aligned} \tag{4.31a}$$

Similarly, here the kinetic term of equation (4.22) becomes

$$\mathcal{L}|_{\text{D-type}} = \Phi_i \Phi^\dagger{}_i\big|_{(\theta\theta)(\bar\theta\bar\theta)}$$
$$= -(\partial_m A^*_i)(\partial^m A_i) + i(\partial_m \bar\psi_i)\bar\sigma^m \psi_i + F^*_i F_i. \quad (4.31b)$$

As was done before, we shall next eliminate the auxiliary fields, using the field equations,

$$\frac{\partial \mathcal{L}}{\partial F^*_i} = F_i + \frac{\partial g}{\partial A_i} = 0.$$

We then have,

$$F_i = -\frac{\partial g}{\partial A_i}. \qquad (4.32)$$

We thus write down the most general Lagrangian as

$$\mathcal{L} = -(\partial_m A^*_i)(\partial^m A_i) + i(\partial_m \bar\psi_i)\bar\sigma^m \psi_i$$
$$- \frac{1}{2}\left[\frac{\partial^2 g}{\partial A_i \partial A_j}(\psi_i \psi_j) + \text{h.c.}\right] - \left(\frac{\partial g}{\partial A_i}\right)^* \left(\frac{\partial g}{\partial A_i}\right). \qquad (4.33)$$

Here the expression for the potential, which is negative of the non-derivative part of the Lagrangian with only scalar fields, is given as

$$V = \left(\frac{\partial g}{\partial A_i}\right)^* \left(\frac{\partial g}{\partial A_i}\right). \qquad (4.34)$$

where as usual summation convention for repeated indices is understood.

4.4.2 Vector Superfields

The construction of the Lagrangian from the vector superfield is a little more complicated, although the same principle of using either the F-type terms or the D-type terms will be adopted. As described in section 4.3, we shall always take the vector field in Wess-Zumino gauge, so that the number of component fields is minimal. The vector superfields, as the name indicates, will be utilised to construct the Lagrangians for the gauge fields which are vector fields in four dimensions. We have already recognised the existence of the vector field $v(x)$ as a component of the vector superfield. It is therefore obvious that we should have the parallel of field strengths for the construction of the Lagrangian. This parallel should be a superfield which is invariant under the gauge transformation considered in equation (4.15), which for the field $v(x)$ is explicitly an abelian gauge transformation as given in one of the equations of (4.17). We shall presently see that the corresponding superfield can be taken as

4.4. Supersymmetric Lagrangian

$$W_\alpha(z) = -\frac{1}{4}(\bar{D}\bar{D})D_\alpha V. \qquad (4.35a)$$

The superfield W_α transforms like a spinor, whereas the earlier superfields Φ or V were transforming as Lorentz scalars. Further, we can easily verify that W_α is a chiral superfield, i.e. it satisfies the equation

$$\bar{D}_{\dot{\beta}} W_\alpha = 0. \qquad (4.35b)$$

This is easily seen since there will be three $\bar{D}$ operators in the left hand side of equation (4.35b), which will automatically make the contribution vanish.

We shall next show that the expression (4.35) is invariant as a superfield when we use the transformation (4.15). This is easily seen when we note that under (4.15)

$$W_\alpha \to -\frac{1}{4}(\bar{D}\bar{D})D_\alpha(V + \Lambda + \Lambda^\dagger)$$
$$= W_\alpha - \frac{1}{4}\bar{D}_{\dot{\beta}}[\bar{D}_{\dot{\beta}}, D_\alpha]\Lambda = W_\alpha, \qquad (4.36)$$

where the appropriate chirality properties of the gauge transformation has been utilised. We shall construct the Lagrangian using the superfield W_α. For this purpose we need to express this superfield in terms of the component fields. Since we recognise that W_α is a chiral superfield, as before it will be convenient to use the variable $y = x + i\theta\sigma\bar{\theta}$ as in equation (4.9) instead of x. Hence in Wess-Zumino gauge we also write the vector superfield V in terms of y, θ and $\bar{\theta}$ as

$$\begin{aligned} V &= -\theta\sigma^m\bar{\theta}v_m(y) + i\,(\theta\theta)\,(\bar{\theta}\bar{\lambda}(y)) - i(\bar{\theta}\bar{\theta})\,(\theta\lambda(y)) \\ &+ \frac{1}{2}\,(\theta\theta)\,(\bar{\theta}\bar{\theta})\,(D(y) + i\,\partial_m\,v^m(y)). \end{aligned} \qquad (4.37)$$

We note that the above is the same as the expression in terms of the variable x in equation (4.14) instead of y, except that we have included an extra term in the last term which cancels the extra contribution from the first term when we make an expansion of y in terms of x. We note that in terms of y, θ, $\bar{\theta}$ parallel to the equations (4.8) become,

$$D_\alpha = \frac{\partial}{\partial\theta^\alpha} + 2i\sigma^m{}_{\alpha\dot{\alpha}}\,\bar{\theta}^{\dot{\alpha}}\frac{\partial}{\partial y^m}, \qquad (4.38a)$$

$$\bar{D}_{\dot{\alpha}} = -\frac{\partial}{\partial\bar{\theta}^{\dot{\alpha}}}. \qquad (4.38b)$$

Hence we easily obtain that

$$\begin{aligned} W_\alpha &= -i\,\lambda_\alpha(y) + \left[\delta_\alpha{}^\beta D(y) - \frac{1}{2}(\sigma^m\bar{\sigma}^n)_\alpha{}^\beta(\,\partial_m\,v_n(y) \right. \\ &\left. -\,\partial_n v_m(y))\right]\theta_\beta + (\theta\theta)\sigma^m{}_{\alpha\dot{\beta}}\,\partial_m\,\bar{\lambda}^{\dot{\beta}}(y). \end{aligned} \qquad (4.39)$$

It is clear that if we wish to construct a supersymmetric Lagrangian from W_α we shall consider F-type terms, the above superfield being a chiral superfield. With some minor algebra, using for example equations like (3.7a) and (3.7b), we first note that

$$W^\alpha W_\alpha \big|_{\theta\theta} = -2i\,\lambda^\alpha \sigma^m{}_{\alpha\dot\alpha}\, \partial_m\,\bar\lambda^{\dot\alpha} - \frac{1}{2}v^{mn}v_{mn} + D^2$$
$$+ \frac{i}{4}\epsilon_{mnpq}\,v^{mn}v^{pq}. \qquad (4.40)$$

In the above, we have substituted

$$v_{mn} = \partial_m\,v_n - \partial_n v_m. \qquad (4.41)$$

It is then easy to see that

$$\mathcal{L} \equiv \frac{1}{4}W^\alpha W_\alpha\Big|_{\theta\theta} + h.c.$$
$$= \frac{1}{2}D^2 - \frac{1}{4}v^{mn}v_{mn} - i\,\lambda^\alpha \sigma^m{}_{\alpha\dot\alpha}\,\partial_m\,\bar\lambda^{\dot\alpha}. \qquad (4.42)$$

As earlier, in writing down (4.42) we have permitted ourselves some integrations in parts corresponding to an action integral. We may note that the vector superfield V has mass dimension zero, the spinorial chiral superfields W_α have mass dimension $(3/2)$, and with the mass dimensions of the θ's being $-(1/2)$, the Lagrangian above has the correct dimension. It is clear that the equation (4.42) does serve the purpose of defining the Lagrangian for massless vector and spin half fields. The auxiliary field D has no kinetic term, and as far as the above Lagrangian is concerned, equations of motion shall yield its value as zero. They will be however relevant when we include interactions. The auxiliary field is needed for the definition of the vector multiplet. We can as earlier derive the supersymmetric transformations in terms of the component fields as

$$\delta_\xi\,v^n{}_m = \frac{1}{2}\Big[\xi\sigma^n\,\partial_m\,\bar\lambda + \bar\xi\bar\sigma^n\,\partial_m\,\lambda - (n \to m, m \to n)\Big], \qquad (4.43a)$$
$$\delta_\xi\,\lambda = i\,\xi\,D + \sigma^{mn}\,\xi\,v_{mn}, \qquad (4.43b)$$
$$\delta_\xi\,D = \partial_m\Big[\bar\xi\bar\sigma^m\lambda - \xi\sigma^m\bar\lambda\Big]. \qquad (4.43c)$$

We note that parallel to equation (4.35) we can also define the antichiral spinorial superfield

$$\bar{W}_{\dot\alpha} = -\frac{1}{4}(DD)\bar{D}_{\dot\alpha}V \qquad (4.44)$$

and use both (4.35) and (4.44) to construct the above Lagrangian. It is also interesting to note that we can also obtain the Lagrangian as a D-type term given as

$$\mathcal{L} = \frac{1}{4}\Big[W^\alpha D_\alpha V + \bar{W}_{\dot\alpha} \bar{D}^{\dot\alpha} V\Big]_{(\theta\theta)(\bar\theta\bar\theta)}, \qquad (4.45)$$

where as earlier we are to do some integration in parts to write the contributions in the conventional form.

In the above we have considered massless vector mesons as well as massless spin half particles. In general we should include a mass term. In such a case we might consider the D-type term arising from $m^2 V^2$ for the Lagrangian, but clearly here we shall not have invariance under the transformation (4.15). With equation (4.14) we note that we have

$$\begin{aligned}V^2\big|_{(\theta\theta)(\bar\theta\bar\theta)} &= \frac{1}{2}v_m v^m - \chi\lambda - \bar\chi\bar\lambda + \frac{1}{2}(M^2 + N^2) \\ &+ \frac{1}{2}\chi\sigma^m\,\partial_m\,\bar\chi + \frac{1}{2}\bar\chi\bar\sigma^m\,\partial_m\,\chi + \frac{1}{2}C\Box C + CD.\end{aligned} \qquad (4.46)$$

We shall however find it more interesting to generate the masses of vector mesons dynamically through spontaneous breaking of local gauge symmetry, and we shall subsequently give many examples of the same.

In the above we have constructed the Lagrangian corresponding to a single vector superfield with no interactions within the vector supermultiplet or with the matter fields corresponding to the chiral multiplets. When we introduce local gauge symmetry corresponding to nonabelian gauge groups and/or including the gauge transformations for the chiral superfields, automatically the interactions in high energy physics as we understand it today, will emerge. We shall consider these in the next chapter.

4.5 Outlook

The present chapter mainly deals with the construction of supersymmetric Lagrangian without including nonabelian gauge theories, which we shall be considering in the next chapter. However, even at this stage some salient features of supersymmetry may be noted.

It is important to note that the potential as given in equation (4.34) is always positive definite. Such a feature was not earlier present in field theory. It leads for example to the hope that such a theory will always lead to a stable vacuum. This feature will subsequently be not there when we consider supergravity, where the positivity of the energy density will no longer be ensured. A further remark in this context is also useful. We know that the cosmological constant of the universe is almost zero. However, the above nonvanishing of the expectation value of the energy density during symmetry breaking automatically leads to a nonvanishing cosmological constant which thus makes supersymmetric theories for the physical world as inappropriate. We shall see however that the above have a lot more complexities associated with the explicit model constructions, and we shall discuss them as and when the appropriate technology or the concept gets developed particularly in the context of supergravity where the above statement gets altered.

REFERENCES

1. A. Salam and J. Strathdee, Nucl. Phys. B 76, 477 (1974); Phys. Lett. 51B, 353 (1974); J. Wess and J. Bagger, *Introduction to Supersymmetry*, Princeton University Press, NJ, 1983.

2. J. Wess and B. Zumino, Nucl. Phys. B 70, 39 (1974); S. Ferrara, B. Zumino and J. Wess, Phys. Lett. 51B, 239 (1974).

Chapter 5

Local Gauge Symmetry

5.1 Introduction

In the last chapter, a supersymmetric invariant Lagrangian was constructed out of chiral as well as with vector superfields. Here we shall proceed to introduce gauge invariance in the Lagrangian [1] which will describe the interaction of vector multiplet with the chiral multiplets along with supersymmetry. As was shown before in equation (4.17) vector multiplet with an abelian gauge transformation contains the vector (gauge) field $v_m(x)$, the spinor (gaugino) field $\lambda(x)$ and an auxiliary field $D(x)$. We shall now generalise this to nonabelian case [2,3] to construct the gauge invariant interactions. In the subsequent sections we shall also introduce in the Lagrangian the F-type and D-type terms as was done before and give a brief note on supersymmetric unification models.

5.2 Gauge Invariant Lagrangian

In the previous chapter, we have seen that the supersymmetric invariant Lagrangian contains (see e.g. equation (4.21)) a kinetic term, the mass term and the interaction term. But when one introduces the gauge transformation, it does not remain invariant and one has to construct it in a gauge covariant manner by introducing a gauge field. In this chapter, we shall take a chiral superfield, which is also a gauge multiplet, and consider the gauge transformation for the same so that for gauge invariance the role of an additional supersymmetric gauge multiplet introduced for the above purpose becomes transparent. We then define the Lagrangian for the gauge multiplet itself. As usual, we shall see that the above two parts of the Lagrangian will separately have gauge invariance.

5.2.1 Lagrangian for Matter Field

Let us now consider a gauge group G. The local gauge transformation for the chiral gauge multiplet $\Phi(y, \theta)$ in the space of superfields is defined as,

$$\Phi \rightarrow \Phi' = e^{-ig\Lambda} \Phi, \qquad (5.1)$$

where $\Lambda = t_a \Lambda_a$ with Λ ($a = 1, 2, \cdots, d$) being the chiral gauge parameters as appropriate functions of space-time and d is the dimension of the gauge group G. t_a's are the generators corresponding to the particular representation of the chiral superfield Φ. Clearly Φ' remains chiral after the gauge transformations since Λ's are classical chiral superfields with $A_a(x), \psi_a(x), F_a(x)$ as the components. To achieve gauge invariance we now replace the kinetic term by,

$$\Phi^\dagger \Phi \Big|_{(\theta\theta)(\bar\theta\bar\theta)} \rightarrow \Phi^\dagger e^{gV} \Phi \Big|_{(\theta\theta)(\bar\theta\bar\theta)}. \qquad (5.2)$$

where $V = V(x, \theta, \bar\theta)$ is a matrix consisting of vector superfield which will contain the supermultiplets corresponding to the gauge particles. In the above we explicitly note that we have taken Φ as a column vector with components Φ_i, automatically $\Phi^\dagger$ is a row vector, and, $V = t_a V^a$ is, as stated, a matrix which defines the d vector superfields for the gauge multiplets. We may also note that in terms of the component fields, $\Phi_i = (\phi_i, \chi_i, h_i)$ with the fields being respectively scalars, spinors and auxiliary fields as described in the last chapter. There is a change in notation here for the components of the chiral superfields since we have taken the superfields Λ_a to have the component fields (A_a, ψ_a, F_a). We note that ϕ and χ are quantum fields, whereas A and ψ are classical fields corresponding to the gauge transformations for the superfields. Further, for the vector superfields, we shall choose Wess-Zumino gauge, so that $V = (v^a{}_m, \lambda^a, D^a)$ with $a = 1, .., d$. The component fields are respectively the spin one gauge fields, spin half gaugino fields, and the auxiliary fields as stated subsequent to equation (4.18) in the last chapter. With equation (5.1) clearly for gauge invariance of the kinetic term in equation (5.2) we shall take the transformations of the vector superfield as,

$$e^{gV} \rightarrow e^{gV'} = e^{-ig\Lambda^\dagger} e^{gV} e^{ig\Lambda}. \qquad (5.3a)$$

For infinitesimal Λ and *an abelian gauge group* this yields that

$$V' = V + i(\Lambda - \Lambda^\dagger), \qquad (5.3b)$$

5.2. Gauge Invariant Lagrangian

which is the parallel of equation (4.15) with a minor change of notation with Λ being replaced by $i\Lambda$. In fact, equation (5.3a) is the nonabelian generalisation of equation (4.15) or of (5.3b). The choice of Wess-Zumino gauge as before shall be always taken. Automatically the chiral superfields Λ shall be restricted so that the vector supermultiplets in Wess-Zumino gauge remain in this gauge after the transformation of equation (5.3a), a property which has been already discussed in chapter 4. With a transformation for the chiral field as in equation (5.1) and with the transformation (5.3a) for the gauge superfield, it is clear that the modified kinetic term for the chiral superfields as in equation (5.2) remains invariant. Further, under the transformation (5.1), the old kinetic term as in equation (4.21a) does not remain invariant, and thus the introduction of the gauge supermultiplet with a specific transformation property as in equation (5.3a) was necessay to maintain the gauge invariance for the kinetic term of the chiral superfield. We shall evaluate the kinetic term (5.2) of the Lagrangian in terms of the component fields. The vector superfield given as in equation (4.14) in Wess-Zumino gauge becomes,

$$V(x,\theta,\bar\theta) = -\theta\sigma^m\bar\theta v_m(x) + i(\theta\theta)(\bar\theta\bar\lambda(x)) - i(\bar\theta\bar\theta)\theta\lambda(x)$$
$$+ \frac{1}{2}(\theta\theta)(\bar\theta\bar\theta)D(x). \qquad (5.4)$$

Now to evaluate $\Phi^\dagger e^{gV} \Phi\big|_{(\theta\theta)(\bar\theta\bar\theta)}$ in terms of component fields, we note that

$$e^{gV} = 1 + gV + \frac{1}{2} g^2 V^2 \qquad (5.5a)$$

where

$$V^2 = \frac{1}{2} (\theta\theta)(\bar\theta\bar\theta) v^{mn}(x) v_{mn}(x). \qquad (5.5b)$$

The other higher powers of V are zero. We also note that for the chiral superfields we shall choose the expansion as in equation (4.13) along with the corresponding equation for the antichiral superfields $\Phi^\dagger$ and then repeatedly use equations (3.7a) and (3.7b) with the parallel equations for products of $\bar\theta$'s and some other equations in the same set. We can then easily collect the coefficients of the terms $(\theta\theta)(\bar\theta\bar\theta)$ in terms of the component fields, where, we are to remember the structure of the component fields as row vectors or column vectors for "matter" fields, and, matrices for the gauge fields. We then have an equation like

$$\Phi^\dagger(x,\theta,\bar\theta) e^{gV} \Phi(x,\theta,\bar\theta)\big|_{(\theta\theta)(\bar\theta\bar\theta)}$$
$$= h^\dagger h + \phi^\dagger \Box \phi + i\ \partial_m\ \bar\chi\bar\sigma^m\chi$$

$$+ \frac{g}{2}\Big[\bar{\chi}\bar{\sigma}^m v_m \chi + i\phi^\dagger v^m \partial_m \phi - i(\partial_m \phi^\dagger)v^m \phi\Big]$$
$$- \frac{i}{\sqrt{2}}g(\bar{\chi}\bar{\lambda}\phi - \phi^\dagger \lambda \chi) + \frac{1}{2}\phi^\dagger(gD + \frac{1}{2}g^2 v^m v_m)\phi. \qquad (5.6a)$$

This equation when rearranged with integration in parts for the action integral, gives the following more suitable form as the gauge invariant generalisation of the kinetic part of the Lagrangian as,

$$\mathcal{L}\big|_{\text{kin}} = h^\dagger h + (D_m \phi)^\dagger (D^m \phi) + i\frac{g}{\sqrt{2}}(\phi^\dagger \lambda \chi - \bar{\chi}\bar{\lambda}\phi)$$
$$- i\bar{\chi}\bar{\sigma}^m D_m \chi + \frac{g}{2}\phi^\dagger D \phi, \qquad (5.6b)$$

where, with $v_m = v^a_m t_a$,

$$D_m = \partial_m + i\frac{g}{2} v_m. \qquad (5.6c)$$

We may note that the spin one gauge bosons corresponding to the nonabelian gauge group as usual interact with the spin half as well as the spin zero matter multiplets through the covariant derivative as in equation (5.6c). Further, there is also an interaction of the above matter multiplets with the spin half gaugino fields. Equations (5.6) thus clearly show the coupling of the supersymmetric matter multiplet to the supersymmetric gauge and gaugino fields characterised by a single coupling constant g. As noted, the chiral supermultiplet is written as $\Phi_i = (\phi_i, \chi_i, h_i)$ so that ϕ, χ and h are column vectors in gauge space. Similarly, v_m, λ and D are matrices in gauge space written in terms of the component fields corresponding to the vector superfields in the adjoint representation of the gauge group G with, as already mentioned, e.g. $v_m = v^a_m t_a$. We may note that the alphabet D is being used with many connotations. D_m is the gauge covariant derivative as in equation (5.6c) with $m=(0,1,2,3)$. In addition D^a stands for the matrix of auxiliary fields of the gauge multiplet with D_a as any one of the d such components. Further, D_α stands for the supersymmetric covariant derivative in superspace giving rise to terms like $DD = D^\alpha D_\alpha$ with the usual summation convention for repeated indices. The context will make the notation transparent, and confusion should not arise.

The second term in equation (6.6b) on the right hand side above is the gauge invariant kinetic term for the scalar fields. The third term there is the interaction term of the gaugino fields λ, the fermionic matter fields χ and scalar fields ϕ. The fourth term is the usual gauge invariant kinetic term for the chiral fermions, and the fifth term gives the interaction of the scalar fields ϕ with the auxiliary gauge fields D.

A few remarks regarding the construction of the gauge invariant kinetic term for supermultiplets may be relevant. Firstly, we may note

5.2. Gauge Invariant Lagrangian

that matter multiplets as considered in particle physics may be either spin half fields, as for quarks and leptons, or, scalar fields, as for the Higgs particles of grand unification type of theories. Hence with supersymmetry it is quite appropriate to take these as chiral supermultiplets with each convenional particle having a superpartner. Next, after a gauge transformation, a chiral supermultiplet as above must remain a chiral supermultiplet so that the gauge transformations be in tune with supersymmetry. Hence the permissible gauge transformations have to be generated with local chiral superfields Λ_a's as is taken in equation (5.1). Usually gauge invariant interactions are introduced by replacing ordinary derivatives by gauge covariant derivatives. However, the kinetic terms for the chiral superfields as in equation (4.21a) of the last chapter does not have any explicit derivative term visible in the space of superfields. This difference in the expression for the kinetic term in terms of superfields thus precludes the method of replacing ordinary derivatives by gauge covariant derivatives for the introduction of local gauge invariance. On the other hand, we recall that local gauge symmetry in ordinary field theory needs the presence of d spin one gauge bosons where d is the dimension of the gauge group G. The simplest supersymmetric generalisation of a spin one boson is that of a vector superfield consisting of the spin one along with spin half component fields as discussed in subsection 4.4.2. Hence the simplest generalisation of gauge bosons with supersymmetry is to take a multiplet of vector superfields. As earlier, these should belong to the adjoint representation of the gauge group. In equation (5.2) such a multiplet has been introduced, with the transformation property as in equation (5.3a) so that the gauge invariance for the modifed kinetic term of equation (5.2) is guaranteed. The choice of Wess-Zumino gauge for the gauge supermultiplets is automatic so that the vector multiplets remain minimal. That the construction of the gauge covariant derivative is not an input for the construction of the gauge invariant kinetic term should not thus surprise us, since, as stated, the construction of the kinetic term itself in supersymmetry in terms of chiral superfields as in equation (4.21a) is formally different.

5.2.2 Lagrangian for gauge field

We shall next proceed to construct the pure gauge Lagrangian of the vector superfield. One has to consider a supersymmetric generalisation of Yang Mills field strength formed from the vector superfields. We now define parallel to the abelian case of the last chapter the chiral spinor "field strength" W_α given as

$$W_\alpha = -\frac{1}{4g}\bar{D}_{\dot\alpha}\bar{D}^{\dot\alpha}\, e^{-gV} D_\alpha\, e^{gV} \qquad (5.7)$$

such that, $\bar{D}_{\dot\alpha} W_\beta = 0$. We recall that in the abelian case we had $W_\alpha = -(1/4)\bar{D}\bar{D}\, D_\alpha V$, and we may verify that the above equation reduces to the same when there is no noncommuting algebra involved. Clearly W_α in equation (5.7) is a matrix in gauge space. It can be written as $W_\alpha = W^a{}_\alpha\, t_a$ corresponding to the d components in adjoint representation for the field strengths as we shall show below.

We next proceed to show that the above matrix also has the same transformation properties as that of field strengths in Yang Mills theory, which will justify the name and will permit us to construct the gauge invariant Lagrangian in a parallel fashion from the present spinor field strengths. For abelian case we had obtained in equation (4.36) that under a gauge transformation,

$$W_\alpha \to -\frac{1}{4}\bar{D}\bar{D}\, D_\alpha\left[V + i\Lambda - i\Lambda^\dagger\right] = W_\alpha, \qquad (5.8)$$

which reflected the fact that the field strengths are invariant under gauge transformations when the gauge group is abelian. For a non-abelian gauge group, from equation (5.7) with (5.3) we get

$$W'_\alpha = -\frac{1}{4g}\bar{D}\bar{D}\, e^{-ig\Lambda}\, e^{-gV}\, e^{ig\Lambda^\dagger} D_\alpha\left[e^{-ig\Lambda^\dagger}\, e^{gV}\, e^{ig\Lambda}\right],$$

which yields that

$$e^{ig\Lambda}\, W'_\alpha = -\frac{1}{4g}(\bar{D}\bar{D}e^{-gV} D_\alpha e^{gV})e^{ig\Lambda} + \frac{1}{4g}\bar{D}\bar{D}D_\alpha\, e^{ig\Lambda}.$$

The second term on the right hand side above is zero. Hence we now obtain the matrix equation

$$W'_\alpha = e^{-ig\Lambda}\, W_\alpha\, e^{ig\Lambda}. \qquad (5.9)$$

Equation (5.9) implies that the chiral superfields W_α transforms as a matrix under nonabelian gauge transformations the same way as the matrix of the Yang-Mills field strengths. The supersymmetric generalisation for the Yang-Mills Lagrangian can therefore now be taken to be in the same form as in equation (4.42) earlier by taking appropriate trace. We shall presently show that

$$\mathcal{L}\big|_{\text{Yang-Mills}} = \frac{1}{2}Tr\left[W^\alpha W_\alpha\big|_{(\theta\theta)} + \bar{W}_{\dot\alpha}\bar{W}^{\dot\alpha}\big|_{(\bar\theta\bar\theta)}\right]. \qquad (5.10)$$

5.2. Gauge Invariant Lagrangian

is the nonabelian generalisation of equation (4.42). By equation (5.9) it is clear that the above Lagrangian is gauge invariant as well as is invariant under supersymmetric transformations modulo four divergence terms. We have to see that it correctly describes the kinetic terms for the spin one gauge bosons and includes the expected interactions of the vector gauge fields with each other, so that usual nonabelian gauge theory is ensured. We also expect that due to supersymmetry it will also include the kinetic terms for the gaugino fields, as well as some fresh interactions of the gaugino fields within the gauge supermultiplet. We shall now evaluate the expression for W_α and explicitly obtain the gauge invariant Lagrangian in terms of the component fields so that we may see the above features. We may first note that W_α is a chiral superfield, and hence as through equations (4.9) and (4.10), it should be a function of the variables $y = x + \theta\sigma\bar{\theta}$ and θ. Hence we shall rewrite first the vector superfield V as a function of y, θ and $\bar{\theta}$. We then get the same equation as (4.37), except that V and thus the component fields are now matrices. With the definition of W_α as in equation (5.7) in mind, we therefore now write the corresponding expressions explicitly in terms of y, θ and $\bar{\theta}$. We first note that, now with $\partial_m = \partial/\partial y^m$,

$$e^{gV} = I - g\,\theta\sigma^m\bar{\theta}\,v_m(y) + ig\,(\theta\theta)\,(\bar{\theta}\bar{\lambda}(y)) - ig(\bar{\theta}\bar{\theta})(\theta\lambda(y))$$
$$+ \frac{1}{2}g\,(\theta\theta)(\bar{\theta}\bar{\theta})\left[D(y) + i\,\partial_m v^m + \frac{1}{2}g\,v^m v_m\right]. \quad (5.11)$$

We also recall that with y, θ and $\bar{\theta}$ as the new variables, the covariant derivatives are given in equations (4.38) of last chapter as

$$D_\alpha = \frac{\partial}{\partial\theta^\alpha} + 2i\sigma^m{}_{\alpha\dot{\alpha}}\bar{\theta}^{\dot{\alpha}}\,\partial_m\;;\quad \bar{D}_{\dot{\alpha}} = -\frac{\partial}{\partial\bar{\theta}^{\dot{\alpha}}}.$$

Further, clearly $\bar{D}\bar{D}(\bar{\theta}\bar{\theta}) = -4$. Hence, in order to evaluate W_α we shall now use that

$$-\frac{1}{4}\bar{D}\bar{D}\left[e^{-gV}D_\alpha e^{gV}\right] = e^{-gV}D_\alpha e^{gV}\big|_{(\bar{\theta}\bar{\theta})}. \quad (5.12)$$

We shall use equation (5.11) and the corresponding expression for e^{-gV} as well as the expression for D_α to simplify the above equation. This simplification is straightforward but cumbersome. We are particularly to remember that the component fields λ, v_m and D are all matrices in group space G and thus they do not commute with each other. We then obtain from equation (5.7) that

$$W_\alpha = -i\lambda_\alpha + \left[\theta_\alpha D - (\sigma^{mn})_\alpha{}^\beta\theta_\beta\left(\partial_m v_n - i\frac{g}{2}(v_m v_n)\right)\right]$$
$$+ (\theta\theta)\sigma^m{}_{\alpha\dot{\alpha}}\left[\partial_m\bar{\lambda}^{\dot{\alpha}} + i\frac{g}{2}(v_m\bar{\lambda}^{\dot{\alpha}} - \bar{\lambda}^{\dot{\alpha}}v_m)\right], \quad (5.13)$$

where,

$$\sigma^{mn} = \frac{i}{2}(\sigma^m \bar{\sigma}^n - \sigma^n \bar{\sigma}^m), \tag{5.14a}$$

such that

$$\sigma^m \bar{\sigma}^n = \eta^{mn} - i\sigma^{mn}. \tag{5.14b}$$

In the above, we have used as earlier that $\theta^\alpha \theta^\beta = -\frac{1}{2}\epsilon^{\alpha\beta}(\theta\theta)$ and that $\bar\theta^{\dot\alpha}\bar\theta^{\dot\beta} = \frac{1}{2}\epsilon^{\dot\alpha\dot\beta}(\bar\theta\bar\theta)$ as well as some other results of chapter 3. The last terms in the two square brackets in equation (5.13) involve commutators of the generators of the group G. Hence through the structure constants of the Lie group, we shall have ultimately all the terms on the right hand side of equation (5.13) expressed in terms of a linear combination of the generators as stated earlier, so that the d components of the spinor field strengths in W_α can be identified.

In equation (5.13), we could also do another simplification. With $D_m v_n = (\partial_m + (ig/2)v_m)v_n$, we may substitute

$$v_{mn} = D_m v_n - D_n v_m \tag{5.15}$$

for the matrix of field strengths for the gauge bosons. We shall presently do so. We then obtain that

$$\begin{aligned}
W^\alpha W_\alpha\big|_{(\theta\theta)} &= -\lambda^\alpha \Big[i\sigma^m{}_{\alpha\dot\alpha} \{ \partial_m \bar\lambda^{\dot\alpha} + (ig/2)(v_m \bar\lambda^{\dot\alpha} - \bar\lambda^{\dot\alpha} v_m) \} \Big] \\
&+ i\sigma^m{}_{\alpha\dot\alpha} \{ \partial_m \bar\lambda^{\dot\alpha} + i\frac{g}{2}(v_m \bar\lambda^{\dot\alpha} - \bar\lambda^{\dot\alpha} v_m) \}\lambda^\alpha + DD \\
&- \frac{1}{2} v_{mn} v^{mn} + \frac{i}{4} \epsilon^{mnlk} v_{mn} v_{lk}.
\end{aligned} \tag{5.16}$$

In deriving the above we have utilised equation (5.15) and that

$$\epsilon^{\alpha\beta}\epsilon_{\gamma\delta}(\sigma^{mn})_\alpha{}^\gamma (\sigma^{lk})_\beta{}^\delta = 2(\eta^{ml}\eta^{nk} - \eta^{mk}\eta^{nl}) - 2i\,\epsilon^{mnlk} \tag{5.17}$$

with $\epsilon^{0123} = 1$. For the calculation of $\bar W_{\dot\alpha} \bar W^{\dot\alpha}$ as required in equation (5.10), we can proceed in a parallel fashion, or take the hermitian conjugate in equation (5.16) to obtain the corresponding contribution.

We shall now proceed to evaluate the Lagrangian as in equation (5.10). For the trace of generators of G in equation (5.16), we take the normalisation that

$$Tr(t_a t_b) = \frac{1}{2}\delta_{ab}. \tag{5.18a}$$

We shall also make use of the gauge covariant derivatives for the gaugino fields with the notation that

$$D_m{}^{ab} = \delta_{ab}\partial_m - \frac{g}{2}f_{cba}v^c{}_m, \tag{5.18b}$$

5.2. Gauge Invariant Lagrangian

where the structure constants f_{abc} of the group G are given by the algebra

$$[t_a, t_b] = i f_{abc} t_c. \tag{5.18c}$$

We may note that in equation (5.18) we shall have $f_{cba} = -f_{abc}$ in case the structure constants are totally antisymmetric. From equations (5.10) and (5.16), on adding appropriate contributions from the antichiral field strengths $\bar{W}_{\dot{\alpha}}$, we then obtain for the supersymmetric nonabelian gauge Lagrangian as

$$\mathcal{L}_{\text{Yang-Mills}} = -i\lambda^{a\alpha}\sigma^m{}_{\alpha\dot{\alpha}} D_m{}^{ab} \bar{\lambda}^{b\dot{\alpha}} - \frac{1}{4} v^a{}_{mn} v^{amn} + \frac{1}{2} D^a D^a. \tag{5.19}$$

As stated, v_{mn} is the Yang-Mills field strength. The second term of equation (5.18) as occuring in (5.19) gives the interaction of the gaugino λ and the vector particle v_m within the supersymmetric gauge multiplet. It may be noted here that for the abelian case, this coupling of the gaugino is absent, just as it is absent amongst the gauge bosons themselves.

We have thus constructed the general Lagrangian for chiral superfields carrying a representation of a nonabelian group G interacting with gauge superfields V. This is invariant under supersymmetry as well as for nonabelian gauge transformations. The total Lagrangian is given as,

$$\begin{aligned}\mathcal{L} &= \Phi^\dagger e^{gV} \Phi\big|_{(\theta\theta)(\bar{\theta}\bar{\theta})} + \mathcal{L}_{\text{Yang-Mills}} \\ &+ \frac{1}{2} m_{ij} \Phi_i \Phi_j \big|_{(\theta\theta)} + g(\Phi)\big|_{(\theta\theta)} + h.c..\end{aligned} \tag{5.20}$$

In the above, $\mathcal{L}_{\text{Yang-Mills}}$ is as in equation (5.10) or (5.19). Further, the mass matrix m_{ij} besides being symmetric must be such that the invariance under the gauge group G for these terms is maintained. The superpotential $g(\Phi)$ in terms of the chiral supermultiplets must also be gauge invariant. Here it may be noted that the Lagrangian contains the F and D terms as in the last chapter. The construction of F-type terms for chiral superfields here remains unaltered in form except for the additional requirement of gauge invariance. The modified D-term as prescribed yields the kinetic term for the component fields of the chiral superfields along with the interaction terms of the same with the component fields of the gauge supermultiplets, which is just the parallel of what was happening earlier when local gauge symmetry was introduced without any supersymmetry. The kinetic terms for gauge bosons and the gauginos along with interactions within the gauge supermultiplets are given by additional F-type terms involving the matrix chiral gauge superfields W_α and $\bar{W}_{\dot\alpha}$ constructed from the gauge vector supermultiplets in V. The parallel of the final contributions as in equation (5.10) and (5.19) with that of a vector supermultiplet as in equations (4.42) and (4.43) of last chapter may be particularly noted.

Here the auxiliary fields D's (for the vector superfields) and h's (for the chiral superfields) are to be eliminated through their respective equations of motion. The elimination of h's remains unaltered as in equation (4.31) giving rise to the F-type part of scalar potential as $(\partial g/\partial \phi_i)^*(\partial g/\partial \phi_i)$ where g is the superpotential *now including the mass term*. For the elimination of the auxiliary gauge fields D's, we note that in the terms for the supersymmetric Yang-mills contribution to the Lagrangian of equation (5.19) we have parallel to equation (4.43) a quadratic contribution of the gauge auxiliary fields. Also, in terms of the chiral superfields, there is a linear contribution from these auxiliary fields as in the last term of equation (5.6b). Hence the elimination of the D's and the calculation of the corresponding contribution to the potential is straightforward. In fact we have from equation (5.19) and (5.6b),

$$D^a = -\frac{g}{2}\, \phi^\dagger\, t_a\, \phi. \qquad (5.21)$$

Substituting this value and adding the F-type and the D-type contributions to the potential, we thus obtain

$$V = \left|\left(\frac{\partial g}{\partial \phi_i}\right)\right|^2 + \frac{g^2}{8}\left|(\phi^\dagger t_a \phi)\right|^2. \qquad (5.22)$$

We shall in later chapters use the above potential, and, the concepts, to construct the supersymmetric unification models as well as consider spontaneous symmetry breaking. For the latter however we shall be considering potentials which shall be generalisations of the expression in (5.22).

5.3 Electroweak Theory

In this section, we shall discuss briefly some guidelines for constructing supersymmetric unification models with electroweak symmetry as a natural example. The basic ingredients for the same are as follows. First we choose a gauge group G with the gauge fields described by vector multiplets $V^a = (v^a, \lambda^a, D^a)$ and then take "matter" sector with chiral superfields belonging to some representations of G. We then construct a G-invariant superpotential g of the chiral superfields, and consider the dynamics that results from the corresponding Lagrangian as

5.3. Electroweak Theory

derived in the last section which will naturally have both gauge symmetry and supersymmetry. We here give an orientation regarding the above with the supersymmetric generalisation of the electroweak unification, which will naturally form the basis for any unification scheme.

For electroweak unification, the gauge group G, denoted as G_{std}, is given as [4] $SU(3)_C \times SU(2)_L \times U(1)_Y$. The corresponding vector particles of the gauge supermultiplet are the eight gluons for the color group $SU(3)_C$, the three W gauge bosons for the weak isospin group $SU(2)_L$, and the B meson for the weak hypercharge group $U(1)_Y$. Thus the corresponding matrix from the vector superfields in group space becomes

$$V = V^a t_a, \qquad (5.23)$$

where, $a = (1,..,8)$ stand for $SU(3)_C$ indices, $a = (9, 10, 11)$ stand for $SU(2)_L$ indices, and, $a = 12$ stands for the $U(1)_Y$ index. We note that at low energies the above gauge symmetry will be broken, $W^{\pm}$ will become massive, and, a mixing between W^3 and B will give rise to a massive vector boson Z and the massless photon. Also, supersymmetry has to be broken so that the predictions be consistent with the experimental fact that the spin half supersymmetric partners of the above gauge particles, i.e. the gauginos, as well as the scalar partners of quarks and leptons have not been yet observed. In the present chapter however we shall be considering the theory with both gauge symmetry and supersymmetry being good in order to illustrate the basic dynamical structure. Symmetry breaking will be considered in later chapters.

The matter chiral multiplet consists of the conventional fermions along with their superpartners as discussed below. We shall also have graviton and gravitino as a supermultiplet in the context of gravity interactions which will be considered later when we discuss supergravity. In the minimal model, it is also usual to introduce two Higgs doublets for the Higgs particles as well as the corresponding Higgsinos. We note that the Higgs particles as well as the Higgsinos are yet to be observed. We have given the particle structure in the context of the standard supersymmetric model in Table 1, with the Gellmann identification of the electromagnetic charge as

$$Q_{em} = I_3 + \frac{1}{2}Y. \qquad (5.24)$$

There are three generations of quarks and leptons which will be designated by $i=1,2,3$ respectively for the generations of (i) e, ν_e leptons and u, d quarks, (ii) μ, ν_μ leptons and c, s quarks, and, (iii) τ, ν_τ leptons and t, b quarks. In the above, the top quark t is not yet seen but we strongly believe from the verified part of the symmetry structure that it will exist. In Table 1 we have mentioned only about the first generation of particles, and the structure of the quarks and leptons of second and third generations are trivial extensions.

We now define the chiral multiplets for the electroweak group given as above. We shall take $Q_i{}^{Am}$ as the chiral superfields corresponding to the left handed quark doublets of the ith generation, with $A=(1,2,3)$ being the color index for color triplets, and $m=(1,2)$ being the weak isospin index for $SU(2)_L$ doublets. Also, the weak hypercharge Y for this supermultiplet is $(1/3)$ as may be seen from equation (5.24). Thus these superfields belong to the representation $(3,2,1/3)$ of G_{std}. The $SU(2)_L$ doublet chiral superfields corresponding to the left handed lepton doublets (ν_e, e^-) in the first generation are now written as $L^i{}_m$ for the ith generation. They all belong to the representation $(1,2,-1)$ of G_{std}. We take the chiral supermultiplets corresponding to the right handed u, d quarks and electron respectively as U, D and E. We may also change the description to taking left-handed antiparticles. With this we replace the above by $u^c{}_L$, $d^c{}_L$ and $e^c{}_L$ as has been taken in Table 1. Both the descriptions are equivalent. Clearly U belongs to the representation $(3,1,4/3)$ of G_{std}, and, similarly, D and E belong respectively to the representations $(3,1,-1/3)$ and $(1,1,2)$ of the same group. The alternative group structure with left-handed antiparticles is given in Table 1. One also takes the chiral superfields correspondng to two Higgs doublets as H^m and $\bar{H}^m$, where these respectively belong to the representations $(1,2,-1)$ and $(1,2,1)$ of G_{std}. Usually two Higgs doublets are taken so that the masses of quarks and leptons can be adjusted with greater freedom so as to agree with observations. The two doublets structure for Higgs particles arises more naturally in supergravity models with "grand unification" as we shall see later.

We then have the supersymmetric and gauge invariant kinetic term for the above chiral superfields as

$$\mathcal{L}_{\text{kin}} = \sum_\Phi \Phi^\dagger e^{gV} \Phi \Big|_{(\theta\theta)(\bar{\theta}\bar{\theta})} . \qquad (5.25)$$

In the above, for the generic chiral supermultiplets Φ, clearly the appropriate representations of the generators are to be taken as stated above when we write the matrix V for the vector superfields for evaluation.

We next take the gauge invariant superpotential as

$$\begin{aligned} g &= \lambda_{ij} \, Q_i{}^{Am} \, \epsilon_{mn} \, \bar{H}^n \, U_j^A + \lambda'_{ij} \, Q_i{}^{Am} \, \epsilon_{mn} \, H^n \, \bar{D}_j^A \\ &+ \lambda''_{ij} L_i{}^m \epsilon_{mn} H^n \bar{E}. \end{aligned} \qquad (5.26)$$

In the above, color invariance is obvious, ϵ_{mn} is introduced to ensure weak isospin invariance, and invariance under the abelian weak hypercharge may be easily seen by adding the hypercharge quantum numbers

5.3. Electroweak Theory

of the multiplets as mentioned. We may note that the observed particles as per strong interactions are not the same as the multiplets for the weak interactions. This fact is ensured by a Kobayashi Maskawa rotation [5] which is here incorporated through the set of parameters λ_{ij}, λ'_{ij} and λ''_{ij}. These parameters link the different generations, and are put phenomenologically. They are related to the masses and mixings of quarks, a theoretical understanding of which is at present an open problem. The parameters being diagonal will make the individual generations stable. We have assumed that for all the generations the Higgs particles are the same, and that the masses of the fermions as generated from the vacuum expectation values of the Higgs particles are different because the λ's, the Yukawa couplings of the different generations, are different. With the parameters as above, we have to diagonalise the mass matrix, so that the "weak" d of the Salam Weinberg theory will not only contain the physical d, but will also have some physical s and physical b as a superposition to be calculated from the above mixing. This will be equivalent to the Kobayashi Maskawa rotation for three generations, parallel to the Cabbibo rotation introduced earlier for two generations [6]. In the context of some later developments, we may note that in order that we may have supersymmetry breaking as well as the weak symmetry breaking simultaneously, we should have another Higgs chiral supermultiplet Y which is a singlet under G_{std}, Corresponding to this we may take an additional contribution to the superpotential as

$$g = \lambda\, Y(H^m \epsilon_{mn} \bar{H}^m - \mu^2), \qquad (5.27)$$

where μ is a mass parameter in the weak scale. The corresponding kinetic term for the chiral supermultiplet Y is also now to be added to the expression in equation (5.25). The above superfield Y may be distinguished from the hypercharge!

In the above contributions, if we associate nonzero vacuum expectation values to the charge neutral scalar components h^0 and $\bar{h}^0$ of the Higgs supermultiplets as stated in Table 1, we shall then have a symmetry breaking of the $SU(2)_L \times U(1)_Y$ subgroup to $U(1)_{em}$ with the Gellmann identification of the charge $Q_{em} = I_3 + \frac{1}{2}Y$. Generating such a vacuum expectation value through a minimisation of the potential for spontaneous symmetry breaking however is more complicated, and will be considered in the next chapter for supersymmetry, and later for supergravity in the context including gravity interactions in our unification attempts.

In the present book our usual emphasis will be on symmetry properties. However, there are deep dynamical questions, associated with supersymmetry, which shall not be touched by us. In the context of the present section, we make the obvious remark that the large number

Table 1. Low energy particle multiplets with their superpartners

Particles		Gauge structure with $G_{std} = SU(3)_C \times SU(2)_L \times U(1)_Y$
Gauge sector:		
spin 1	spin 1/2	
gluons	gluinos	$SU(3)_C$
W's	Winos	$SU(2)_L$
B	Bino	$U(1)_Y$
Matter sector:		
spin 1/2	spin 0	
$L = \begin{pmatrix} \mu_\nu \\ e^- \end{pmatrix}_L$	scalar partners	$(1, 2, -1)$
$q = \begin{pmatrix} u \\ d \end{pmatrix}_L$	scalar partners	$(3, 2, 1/3)$
$d^c{}_L$	scalar partners	$(3^*, 1, 2/3)$
$u^c{}_L$	scalar partners	$(3^*, 1, -4/3)$
$e^c{}_L$	scalar partners	$(1, 1, 2)$
Higgs sector:		
spin 0	spin 1/2	
$\bar{H} = \begin{pmatrix} h^+ \\ \bar{h}^0 \end{pmatrix}$	spin 1/2 partners	$(1, 2, 1)$
$H = \begin{pmatrix} h^0 \\ h^- \end{pmatrix}$	spin 1/2 partners	$(1, 2, -1)$
Gravity sector:		
spin 2	spin 3/2	
graviton	gravitino	$(1, 1, 0)$

of Yukawa couplings in equation (5.26) make the superpotential ugly. Besides any possible symmetry structure, it is quite possible that the dynamics of chiral symmetry breaking could enable us to calculate some of them in terms of others. As stated, consideration of these from some deeper principle remains as an open problem. The other feature which is artificial is the peculiar way in which the multiplet structures for quarks and leptons are written down. This however has a reasonable understanding in terms of any of the grand unification theories, as we shall discuss. One starts here with a larger group G, and G_{std} emerges as a result of spontaneous symmetry breaking at scales much higher than

5.3. Electroweak Theory

the weak scale, which is only 100 GeV. The larger scales involved may be any thing from 10^{11} GeV to 10^{15} GeV, or even the Planck scale of 10^{19} GeV. It is believed that at reasonably high energies supersymmetry will be good, and we have not seen any supersymmetric partners because we are at present hardly reaching an energy scale of the order of a TeV! The appeal for supersymmetric theories at these high scales is from the aesthetic beauty of unification as well as from some effects which have consequences at lower energy scales. They should be observed in future experiments.

REFERENCES

1. J. Wess and B. Zumino, Nucl. Phys. B78, 1 (1974)

2. S. Ferrara and B. Zumino, Nucl. Phys. B79, 413 (1974)

3. A. Salam and J. Strathdee, Phys. Lett. 51B, 353 (1974)

4. S.L. Glashow, Nucl. Phys. 22, 579 (1961); S. Weinberg, Phys. Rev. Lett. 19, 1264 (1967); A. Salam, *Elementary Particle Theory*, edited by N. Svartholm) Almquist and Forlag, Stockholm,1968.

5. K. Kobayashi and T. Maskawa, Prog. Theor. Phys. 49, 652 (1973).

6. S.L. Glashow, J. Illiopoulos and L. Mariani, Phys. Rev. D2, 1285 (1970).

Chapter 6

Symmetry Breaking

6.1 Introduction

In the previous chapters, we have illustrated supersymmetry as a boson-fermion symmetry and have incorporated within it local gauge symmetry. However, supersymmetry has to be "badly" broken, since no supersymmetric partner has been observed so far, with the lower bounds for the masses of such particles varying between say 50 GeV to 100 GeV. We shall consider therefore the modality for the breaking of supersymmetry now.

In this chapter we shall consider mostly breaking of supersymmetry by hand with simple examples for the sake of illustrating a point. Further, since it is a global symmetry, spontaneous breaking of such a symmetry will imply the existence of zero mass "Goldstino" mode corresponding to fermionic increments. We shall also consider gauge symmetry breaking with broken supersymmetry through super Higgs mechanism. In the later chapters we shall include spontaneous breaking of supersymmetry through introduction of supergravity, the local version of supersymmetry.

At the outset we may state that such symmetry breaking theoretically may be of three categories. (i) We may have broken supersymmetry with gauge symmetry remaining good; (ii) We may have broken gauge symmetry with supersymmetry remaining good; (iii) We may have both supersymmetry and gauge symmetry broken. Here the spontaneous breakaing of ordinary gauge symmetry is conventional, and, the condition for supersymmetry breaking needs some additional constraints, and will be now examined. We know that, the supersymmetric Hamiltonian operator is given as,

$$H = \frac{1}{4}\left[\bar{Q}_1 Q_1 + Q_1 \bar{Q}_1 + \bar{Q}_2 Q_2 + Q_2 \bar{Q}_2\right], \qquad (6.1)$$

6.2. O'Raifeartaigh Mechanism

where Q_1, Q_2 denote the supersymmetric charge. Thus a necessary and sufficient condition for supersymmetry to be good is that vacuum should have zero energy, where we are using the fact that invariance of vacuum under a group of transformations is equivalent to the corresponding symmetry being unbroken. Further we know that, the scalar potential V of the Hamiltonian is given by,

$$V = FF^* + \frac{1}{2}D^2. \qquad (6.2)$$

Hence, the energy of the vacuum will be nonvanishing when one of the auxiliary fields has nonvanishing vacuum expectation value (vev) so that supersymmetry is spontaneously broken. And the fermionic partner of the auxiliary field that receives a vev is the massless Goldstone fermion of broken supersymmetry.

In the present chapter, we shall deal with some examples of spontaneously broken supersymmetry e.g. the O' Raifeartaigh mechanism and the Fayet-Illiopoulous machanism.

6.2 O'Raifeartaigh Mechanism

In this model [1] the auxiliary field F of the chiral superfield attains vev to break the supersymmetry. This model contains three chiral superfields Φ_0, Φ_1, Φ_2 and the interaction is given by the superpotential

$$g = \lambda\Phi_0 + g\Phi_0\Phi_1\Phi_2 + m\Phi_1\Phi_2. \qquad (6.3)$$

Here we may note that the superpotential has an additional U(1) symmetry under which $\Phi_0 \to \Phi_0$, $\Phi_1 \to -\Phi_1$, $\Phi_2 \to -\Phi_2$. We shall now evaluate the F-terms of the above three chiral fields,

$$F_0^* = \left(\frac{\partial g}{\partial A_0}\right) = \lambda + gA_1A_2, \qquad (6.4a)$$

$$F_1^* = \left(\frac{\partial g}{\partial A_1}\right) = gA_2A_0 + mA_2, \qquad (6.4b)$$

$$F_2^* = \left(\frac{\partial g}{\partial A_2}\right) = gA_1A_0 + mA_1. \qquad (6.4c)$$

Thus the scalar potential corresponding to F-type term is,

$$V = \sum_i |F_i|^2 = |(\lambda + gA_1A_2)|^2 + |(gA_2A_0 + mA_2)|^2$$
$$+ |(gA_1A_0 + mA_1)|^2, \qquad (6.5)$$

which is obviously positive definite. For, the potential vanishes only when the conditions

$$0 = \lambda + gA_1A_2 = gA_2A_0 + mA_2 = gA_1A_0 + mA_1. \quad (6.6)$$

are simultaneously satisfied. There is no solution to (6.6), and thus the potential never vanishes, so that, supersymmetry is broken spontaneously. Thus this model has a broken super- symmetry, although we start with a symmetric theory. If A_0, A_1, and A_2 have all zero vacuum expectation values, then $<F_0> = \lambda \neq 0$ and thus equation (4.12b) for the transformation of ψ_0 under a supersymmetry transformation shows that this is the Goldstino mode for the symmetry breaking.

This is an example of F-type symmetry breaking. We shall mainly consider this type of symmetry breaking in later chapters for supergravity.

6.3 Fayet-Illiopoulos Mechanism

This is a model with D-type symmetry breaking first suggested by Fayet and Illiopoulos [2]. They showed that supersymmetry can be broken spontaneously by adding a supersymmetric and gauge invariant term to the Lagrangian. The supersymmetric Lagrangian with an $U(1)$ gauge symmetry is a particular case of nonabelian gauge symmetry considered in the last chapter. To obtain this we first note that as per equation (5.18a) for the normalisation of the generators, we shall take $t_a = 1/\sqrt{2}$, and correspondingly take $g/\sqrt{2} = e$. Also we have automatically the replacements $v_m \to (1/\sqrt{2})v_m$ and $\lambda \to (1/\sqrt{2})\lambda$. Alternatively we may also obtain the Lagrangian for the $U(1)$ symmetry for the Fayet-Illiopoulos model from the expression

$$\begin{aligned}\mathcal{L} &= \frac{1}{4}\Big[W^\alpha W_\alpha\big|_{(\theta\theta)} + \bar{W}_{\dot\alpha}\bar{W}^{\dot\alpha}\big|_{(\bar\theta\bar\theta)}\Big] + m\Big[\Phi_1\Phi_2\big|_{(\theta\theta)} + \Phi^\dagger_1\Phi^\dagger_2\big|_{(\bar\theta\bar\theta)}\Big] \\ &+ \Big[\Phi^\dagger_1 e^{eV}\Phi_1 + \Phi^\dagger_2 e^{-eV}\Phi_2\Big]_{(\theta\theta)(\bar\theta\bar\theta)} + 2\kappa V\big|_{(\theta\theta)(\bar\theta\bar\theta)}. \quad (6.7)\end{aligned}$$

Here we have coupled two matter chiral superfields Φ_1, Φ_2 to the gauge superfield V. The additional term of the Lagrangian is the gauge invariant D-term which is the $(\theta\theta)(\bar\theta\bar\theta)$ component of the vector superfield V. The scalar potential is,

$$V = \frac{1}{2}D^2 + F_1 F_1^* + F_2 F_2^*, \quad (6.8)$$

where, F_1 and F_2 are determined from the Euler equations,

$$D + \kappa + \frac{e}{2}(A_1 A_1^* - A_2 A_2^*) = 0, \quad (6.9a)$$

$$F_1 + mA_2^* = 0, \quad (6.9b)$$

6.3. Fayet-Illiopoulos Mechanism

$$F_2 + mA_1^* = 0. \tag{6.9c}$$

These three equations obviously imply that the solution with $D = 0$, $F_1 = 0$, $F_2 = 0$ is *impossible* which means that, vacuum energy is nonvanishing and hence supersymmetry is spontaneously broken. But the Lagrangian has an $U(1)$ symmetry initially. Let us now examine the fate of gauge symmetry. We now substitute the solutions F_1, F_2 and D from equation (6.9) in the potential to obtain,

$$\begin{aligned} V &= \frac{1}{2}\kappa^2 + (m^2 + \frac{1}{2}e\kappa)A_1 A_1^* + (m^2 - \frac{1}{2}e\kappa)A_2 A_2^* \\ &+ \frac{1}{8}e^2(A_1 A_1^* - A_2 A_2^*)^2. \end{aligned} \tag{6.10a}$$

We now examine different possibilities such as (i) $m^2 > \frac{1}{2}e\kappa$ and (ii) $m^2 < \frac{1}{2}e\kappa$.

Case (i): $m^2 > \frac{1}{2}e\kappa$.

Clearly for the potential (6.10a) the vacuum expectation values of the fields A_1 and A_2 are zero and the energy density of vacuum is given as $V_{\min} = \frac{1}{2}\kappa^2$. Thus supersymmetry is broken. In fact this describes the two scalar fields A_1 and A_2 with masses m_1 and m_2 such that $m_1^2 = m^2 + \frac{1}{2}e\kappa$ and $m_2^2 = m^2 - \frac{1}{2}e\kappa$, and the two spinor fields, ψ_1 and ψ_2 each with mass m. The gaugino and the vector fields given by λ and v_m remain massless. This can easily be verified when we obtain the Lagrangian as

$$\begin{aligned} \mathcal{L} &= (D_m A_i)^*(D^m A_i) - i\bar\psi_i \bar\sigma^m D_m \psi_i - \frac{1}{4}v^{mn}v_{mn} - i\bar\lambda \bar\sigma^m \partial_m \lambda \\ &+ i\frac{e}{\sqrt{2}}\Big[A_1^*(\lambda\psi_1) - A_2^*(\lambda\psi_2) - h.c.\Big] - m(\psi_1\psi_2 + \bar\psi_1\bar\psi_2) - V. \end{aligned} \tag{6.10b}$$

In the above $i=1,2$ stands for the two superfields, and $D_m = \partial_m + i(e/2)v_m$ for F_1 with $i = 1$ or 2.

We thus conclude that the gauge symmetry remains unbroken with broken supersymetry. The spinor λ is the massless Goldstone fermion as may be seen from the fact that under supersymmetry,

$$\delta\lambda = i\xi D + \sigma^{mn}\xi v_{mn}, \tag{6.10c}$$

and from equation (6.9a) D has a nonzero vacuum expectation value.

Case (ii): $m^2 < \frac{1}{2}e\kappa$.

For this, we shall see that the minimisation equations will yield a consistent nonzero solution for the scalars A_1 and A_2. These equations are given as,

$$\left(\frac{\partial V}{\partial A_1^*}\right) = (m^2 + \frac{1}{2}e\kappa)A_1 + \frac{e^2}{4}(A_1^*A_1 - A_2^*A_2)A_1 = 0, \quad (6.11a)$$

and,

$$\left(\frac{\partial V}{\partial A_2^*}\right) = (m^2 - \frac{1}{2}e\kappa)A_2 - \frac{e^2}{4}(A_1^*A_1 - A_2^*A_2)A_2 = 0. \quad (6.11b)$$

These two equations when solved simultaneously, give $A_1 = 0$, $A_2 = v$, where $\frac{v^2e^2}{4} + m^2 - \frac{1}{2}e\kappa = 0$ with $m^2 < \frac{1}{2}e\kappa$. But we know that, a charged scalar field having a nonvanishing vacuum expectation value induces gauge symmetry breaking. In the present case, the $U(1)$ gauge symmetry is spontaneously broken along with the breaking of supersymmetry. The conclusion can also be verified by explicit mass calculation of the vector meson and the spinors. In fact, the quadratic part of the Lagrangian for the bosonic fields with the substitution $A_1 = A$, $A_2 = \tilde{A} + v$ at the minimum of the potential is given as

$$V = \frac{2m^2}{e^2}(e\kappa - m^2) + 2m^2 A^*A + \frac{1}{4}e^2v^2(\tilde{A} + \tilde{A}^*)^2/2$$
$$+ \frac{e^2v^2}{4}(v^m v_m + \cdots), \quad (6.12a)$$

where

$$v = \frac{2}{e}(\frac{1}{2}e\kappa - m^2)^{1/2}, \quad (6.12b)$$

as obtained from the minimisation. As usual, the mass term of the vector meson in equation (6.12a) arises from the first term on the right hand side of equation (6.10b). Here since $m^2 < \frac{1}{2}e\kappa$, the term $\frac{2m^2}{e^2}(e\kappa - m^2)$ is positive and is less than $\kappa/2$ so that at the minimum of the potential A_2 is given by equation (6.12b). Thus gauge symmetry is broken, and, the vacuum energy being positive definite as earlier also implies that supersymmetry is broken. The squares of the masses of A, $(\tilde{A} + \tilde{A}^*)/\sqrt{2}$, and v_m are respectively $2m^2$, $\frac{1}{2}e^2v^2$, and $2 \times \frac{1}{2}e^2v^2$. The vector field v_m acquires a mass by "eating" the Goldstone boson field $(\tilde{A} - \tilde{A}^*)/\sqrt{2}$ which decouples in the Lagrangian. The fermions have a mixing, and, after mixing, the masses for spinors ψ, $\tilde{\psi}$, $\tilde{\lambda}$ as eigenstates of mass matrix can be calculated. Here $\tilde{\lambda}$ remains as the massless Goldstone fermion. In fact, when we write down the quadratic terms in fermion space parallel to equation (6.12a) we have explicitly the mixings given as

$$\psi = \psi_2, \tag{6.12c}$$

$$\tilde{\psi} = \frac{1}{\sqrt{m^2 + e^2v^2/2}} \left[m\psi_1 + i\frac{ev}{\sqrt{2}}\lambda \right], \tag{6.12d}$$

and

$$\tilde{\lambda} = \frac{1}{\sqrt{m^2 + e^2v^2/2}} \left[m\lambda + i\frac{ev}{\sqrt{2}}\psi_1 \right]. \tag{6.12e}$$

In fact, after diagonalisation in terms of the rotated fields the mass term of fermions in the Lagrangian, we get the same as

$$-(m^2 + e^2v^2/2)^{1/2} \left[\psi\tilde{\psi} + \bar{\psi}\bar{\tilde{\psi}} \right], \tag{6.12f}$$

which shows that ψ and $\tilde{\psi}$ both have mass $(m^2 + e^2v^2/2)^{1/2}$ and, as stated, $\tilde{\lambda}$ remains massless.

We may note that the sum of the squares of the masses of the bosons in case (i) and the same sum for the fermions here both equal $2m^2$. In case (ii), also, the sum of the squares of masses of bosons and separately of fermions are both equal to $2m^2 + e^2v^2$. This is a general property for supersymmetry even when symmetry is broken.

We may also have the possibility of gauge symmetry breaking alone with supersymmetry being good. We shall investigate this possibility in the next section. We shall also consider the Fayet Illiopoulos phase transition again through coherent state construction as vacuum destabilisation in quantum field theory in section (6.6) in this Chapter, where, contrary to conventional classical picture with a background, perturbative and nonperturbative phases shall be related by a unitary taransformation.

6.4 Supersymmetric SU(5) Model

In Chapter 5 we have already discussed the basic ingredients of constructing grand unification models with supersymmetry along with local gauge symmetry. Attempts in this direction were made e.g. in Ref.[3-5]. Here we shall discuss however the explicit attempts in $SU(5)$ based models given by Georgi and Dimopoulos [3] and by Sakai [4]. They considered the symmetry breaking channel

$$SU(5) \supset SU(3)_C \times SU(2)_L \times U(1)_Y \supset SU(3)_C \times U(1)_{em}$$

with supersymmetry being good upto the low energy sector. The corresponding superpotential is given as,

$$\begin{aligned}
g &= \lambda_1 \left[\frac{1}{3} \Sigma_y^{\ x} \Sigma_z^{\ y} \Sigma_x^{\ z} + \frac{1}{2} m \Sigma_y^{\ x} \Sigma_x^{\ y} \right] \\
&+ \lambda_2 H'_x \left(\Sigma_y^{\ x} + 3m' \delta_y^{\ x} \right) H^y \\
&+ f_{jk} \epsilon_{uvwxy} H^u M_j^{vw} M_k^{xy} + g_{jk} H'_x M_j^{xy} M_{ky}.
\end{aligned} \qquad (6.13)$$

In the above the Higgs sector contains $\Sigma_x^{\ y} \in$ 24-plet of $SU(5)$ which breaks $SU(5)$ to $SU(3)_C \times SU(2)_L \times U(1)_Y$ and $H'_x \in \{5^*\}$, $H^y \in \{5\}$ are responsible for the next stage of symmetry breaking given as $SU(2)_L \times U(1)_Y \to U(1)_{em}$. The matter sector consists of M_j^{vw} as the jth generation 10-plet and M_{ky} as k^{th} generation 5^*-plet of $SU(5)$. The possible mixing of the generations through Higgs particles is included here directly through the matrices f_{jk} and g_{jk} which are usually put by hand to agree with the experimental results regarding the quark masses and their mixing to yield the Kobayashi-Maskawa matrix. It may be noted that in supersymmetry we have needed both 5 and 5^* of Higgs multiplets for low energy symmetry breaking for anomaly cancellation which is a requirement for the renormalisability of a theory. For the vacuum configurations of the Higgs particles, the D-terms vanish since the 24-plet representation is real and we have added 5 and 5^* representations which are complex conjugates of each other. We shall thus minimise the scalar potential $V = \sum \left| \left(\frac{\partial g}{\partial \phi} \right) \right|^2$. This yields three sets of solutions given as,

$$< H^x > = < H'^x > = < M_j^{xy} > = < M_{jx} > = 0, \qquad (6.14a)$$

and

$$< \Sigma_y^{\ x} > = \begin{cases} 0, \\ \frac{m}{3} \delta_y^{\ x} - \frac{5}{3} m \, \delta_5^{\ x} \delta_y^{\ 5}, \\ 2m \delta_y^{\ x} - 5m (\delta_4^{\ x} \delta_y^{\ 4} + \delta_5^{\ x} \delta_y^{\ 5}) \end{cases} \qquad (6.14b)$$

Since for all the above configurations potential vanishes, supersymmetry remains good. The first solution does not break the gauge symmetry $SU(5)$, the second breaks it to $SU(4) \times U(1)$ and the third one to $SU(3)_C \times SU(2)_L \times U(1)_Y$ which is the relevant solution for grand unification in the present case. This solution in (6.14) implies the Higgs doublets $H^{x=4,5}$ and $H'_{y=4,5}$ have a mass squared proportional to $(m - m')$. We shall take $m = m'$ in equation (6.13), so that the $SU(2)_L$ doublets remain massless and the color triplet components acquire a mass of the order of m. This doublet triplet mass splitting of the two quintets of Higgs particles is phenomenologically necessary so that proton decay does not take place through the color triplet of Higgs particles at the

6.5. Low energy sector

weak scale, which would violently contradict the observed stability of the proton. In the low energy sector, for the Salam Weinberg group $SU(3)_C \times SU(2)_L \times U(1)_Y$ the symmetry breaking to $SU(3)_C \times U(1)_{em}$ takes place through the nonvanishing vevs of the $SU(2)_L$ doublets in H and H'. At this stage we add soft supersymmetry breaking terms [6] to the Lagrangian, so that the weak symmetry and supersymmetry breaking occur simultaneously at the weak scale. We may now quote some of the relevant soft symmetry breakaing terms given as [6], (i) $m^2(\phi^2 + \phi^{*2})$, (ii) $m^2 \phi\phi^*$; (iii) trilinear scalar operators, (iv) gaugino mass term. Here ϕ denotes a complex scalar field corresponding to the spin zero part of the chiral multiplets. The presence of the above type of terms in the Lagrangian breaks supersymmetry, but does not introduce any additional divergences in the context of renormalisation. This is the reason why they are called soft symmetry breaking terms. In the next subsection, we shall investigate one of these possibilities for the low energy sector.

6.5 Low energy sector

Here we shall consider a model with minimal particle content in which the Higgs sector consists only of H and H'. We shall break supersymmetry by hand, but shall take a form which shall have relevance later in the context of supergravity described through superpotentials. With this in mind, for a given superpotential g, we shall take the potential along with additional soft symmetry breaking terms as

$$V = \left|\frac{\partial g}{\partial \phi}\right|^2 + m_{3/2}^2 |\phi|^2 + m_{3/2}\left[\left(\frac{\partial g}{\partial \phi}\right)\phi - \sqrt{3}g + h.c.\right]. \qquad (6.15)$$

The second and the third terms above are the additional terms which violate supersymmetry. $m_{3/2}$ is a mass parameter in the weak scale, and, as usual, summation over the scalar fields ϕ is understood. Also, in equation (6.15), we have not included the D-term coming from the gauge symmetry as discussed in the last Chapter, but rather wish to emphasize the type of soft symmetry breaking terms taken. For a superpotential which is at best trilinear, the second and third terms above are in fact the soft symmetry breaking terms since they constitute operators of dimension less than four and thus do not permit the renormalisation results to be disturbed. As stated the justification for terms like (6.15) will be on the basis of supergravity and shall be considered in Chapter 10; at present we are only concerned with the same as an example for supersymmetry breaking with soft symmetry breaking terms. Let us now slightly change the notations of the last subsection to take H and H' as the doublets of Higgs particles of the last section, with the triplet sector dropping out by acquiring a mass in the grand unification scale as described in the last subsection.

After repeated trials it was recognised a long time back that in terms of only H and H' it is not possible to break gauge symmetry and supersymmetry simultaneously at the tree level. An additonal singlet with the vev around the weak scale is always needed. To illustrate this, we shall therefore in addition take a chiral superfield Y which is a singlet under the Salam Weinberg gauge group $SU(3)_C \times SU(2)_L \times U(1)_Y$. We also choose a simple superpotential as [7,8]

$$g = \lambda Y(HH' - \mu^2). \tag{6.16}$$

In the above λ is a dimensionless parameter and μ is a mass parameter in the weak scale. Here we have only included the part of the superpotential which is relevant regarding symmetry breaking, and omitted writing down the matter part of it or the part that will give rise to the Yukawa couplings. From equation (6.16), the form of the potential now as in equation (6.15) becomes

$$V = \lambda^2 \left[|hh' - \mu^2|^2 + |yh|^2 + |yh'|^2 \right] + m_{3/2}^2 \left[|y|^2 + |h|^2 + |h'|^2 \right]$$
$$+ m_{3/2} \lambda \left[(3 - \sqrt{3}).yhh' - (1 - \sqrt{3})\mu^2 y + h.c. \right] \tag{6.17}$$

In the above we have denoted by lower case alphabets the scalar components of the corresponding superfields. Thus, y is the complex scalar field corresponding to Y and is a singlet in Salam Weinberg space, and, h and h' are the scalar Higgs doublets corresponding to H and H'. Also, the appropriate scalar products for the different components of the above doublets are understood. The second and third terms in the right hand side of equation (6.17) break supersummetry. When $\mu > \frac{m_{3/2}}{\lambda}$, we obtain a minimum of the potential for the expectation values given as

$$<y> = m_{3/2} \frac{q}{\lambda}; \quad <h> = \begin{bmatrix} \left(\frac{\mu p}{\lambda}\right) \\ 0 \end{bmatrix}; \quad <h'> = \begin{bmatrix} 0 \\ \left(\frac{\mu p}{\lambda}\right) \end{bmatrix}, \tag{6.18}$$

where p and q satisfy the simultaneous equations

$$(2p^2 + 1)q + (\sqrt{3} - 1)[\sqrt{3}p^2 + \mu^2\lambda^2/(m_{3/2}^2)] = 0, \tag{6.19a}$$

$$q^2 + (3 - \sqrt{3})q + 1 + p^2 - \mu^2\lambda^2/(m_{3/2}^2) = 0. \tag{6.19b}$$

The above equations reveal the action of soft supersymmetry breaking terms illustrating the symmetry breaking at the weak scale. We also note that $m_{3/2}$ really stands for the mass parameter corresponding to gravitino as mentioned in Table 1 of Chapter 5. This interpretation however is not relevant as far as the derivations in the present section are concerned. We note that until the weak scale is reached, as discussed in the last subsection, supersymmetry is good. Hence the nonrenormalisation theorems (stated in the next Chapter) will ensure that the renormalisation effects in the grand unification scale will not vitiate the results of the weak scale. This stability of the hierarchy was initially the reason for supersymmetry as such and whether supersymmetry is broken by hand or is spontaneously broken through local supersymmetry at the lower scale is not material to guarantee the above property.

We would like to mention that equations (6.19) are the exact prallel of the equations (10.10) of Chapter 10 where p and q may be respectively replaced by x and y. In Chapter 10 we have considered grand unification with local supersymmetry or supergravity, and symmetry breaking occurs spontaneously through super Higgs mechanism. Within the scope of supersymmetry, we thus wished to have an illustration which is linked with gravity in a deeper manner to be described later.

6.6 Quantum Phase Transition

In the earlier sections we have considered symmetry breaking with the picture that a scalar field may have a classical background value around which the quantum excitations are to be analysed, and the nonzero background field while minimising the potential gives rise to symmetry breaking. We shall reconsider the same in quantum field theory as an example of destabilisation of the vacuum state.

To fix our ideas we illustrate this for the Fayet-Illiopoulos model considered earlier. Let us take a state

$$|vac'> = U|vac> \qquad (6.20)$$

where

$$U = exp(B^\dagger - B), \qquad (6.21)$$

with $B^\dagger$ being some trial operator defined in terms of the scalar fields. We shall construct the Hilbert space for the Higgs particles including interaction. Further, we shall try to determine the operator B of equation (6.21) through a variational analysis and examine whether the energy density for $|vac'>$ is smaller than that for $|vac>$. Now, at time $t=0$, let us expand the two complex scalar fields as

$$A_i(x) = \frac{1}{(2\pi)^{3/2}} \int d^3\vec{k} \frac{1}{\sqrt{2w_i(\vec{k})}} \left[b_i(\vec{k})^\dagger e^{-i\vec{k}\cdot\vec{x}} + a_i(\vec{k}) e^{i\vec{k}\cdot\vec{x}} \right],$$

(6.22a)

and, for its time derivative,

$$\dot{A}_i(x) = \frac{i}{(2\pi)^{3/2}} \int d^3\vec{k} \sqrt{\frac{w_i(\vec{k})}{2}} \left[b_i(\vec{k})^\dagger e^{-i\vec{k}\cdot\vec{x}} - a_i(\vec{k}) e^{i\vec{k}\cdot\vec{x}} \right], \quad (6.22b)$$

Here as usual

$$[a_i(\vec{k}), a_j(\vec{k}')^\dagger] = \delta_{ij}\delta(\vec{k}-\vec{k}'), \qquad (6.23a)$$

$$[b_i(\vec{k}), b_j(\vec{k}')^\dagger] = \delta_{ij}\delta(\vec{k}-\vec{k}'). \qquad (6.23b)$$

Further $w_i(\vec{k}) = (\vec{k}^2 + m_i^2)^{1/2}$ for free fields; here it can be arbitrary with the assumption that it commutes with the above creation and annihilation operators. Then, the equal time algebra for the interacting fields $A_i(\vec{x})$ and $\dot{A}_i(\vec{x}')$ is automatically satisfied as expected. We define $|vac>$ with the requirement that $a_i(\vec{k})|vac> = b_i(\vec{k})|vac> = 0$ and call this the perturbative vacuum.

We now consider the operator

$$B_i{}^\dagger = -\int d\vec{x} v_i(\vec{x}) \sqrt{\frac{w_{i\vec{x}}}{2}} \left[a_i(\vec{x})^\dagger + b_i(\vec{x})^\dagger \right]. \qquad (6.24)$$

In the above, $w_{i\vec{x}} = (-\vec{\nabla}_{\vec{x}}^2 + m_i^2)^{1/2}$ for free field quantisation, and, for the sake of simplicity, we shall take $v_i(\vec{x})$ as real. From the field quantisation we then easily have e.g. without any summation,

$$[A_i(\vec{x}), B_j{}^\dagger] = -\frac{1}{2}\delta_{ij}v_j(\vec{x}). \qquad (6.25)$$

Now, in equation (6.21) we take

$$B^\dagger = B_1{}^\dagger + B_2{}^\dagger. \qquad (6.26)$$

Thus for the definition of $|vac'>$ in equation (6.20), $v_i(x)$ are the trial functions. We now obtain that

$$[A_i(\vec{x}), B^\dagger - B] = -v_i(\vec{x}). \qquad (6.27)$$

Let us now define for a real λ the expression

6.6. Quantum Phase Transition

$$f(\lambda) = e^{-\lambda(B^\dagger - B)} A_i(\vec{x}) e^{\lambda(B^\dagger - B)}.$$

Hence we easily obtain that $df(\lambda)/d\lambda = -v_i(\vec{x})$, so that, integrating and substituting $\lambda = 0$, we get,

$$A_i(\vec{x})' \equiv U^{-1} A_i(\vec{x}) U = -v_i(\vec{x}) + A_i(\vec{x}), \qquad (6.28a)$$

or,

$$A_i(\vec{x}) = v_i(\vec{x}) + A_i(\vec{x})'. \qquad (6.28b)$$

Hence, we may replace $A_i(\vec{x})$ by the quantum field $A_i(\vec{x})'$ as in equation (6.28b) along with a "translation" of the field through a classical expression $v_i(\vec{x})$. This as we see is equivalent to the replacement of $|vac>$ by $|vac'> = U|vac>$. With the calculation of the energy density, we shall find whether $|vac>$ is stable or it destabilises to a $|vac'>$ different from $|vac>$.

In field theory we wish to have translational invariance, and hence we shall take $v_i(\vec{x})$ as independent of $\vec{x}$. We easily see that with V given by equation (6.8), and with equation (6.9) for the auxiliary fields, we have

$$<vac'|V|vac'> = \frac{1}{2}(\kappa + \frac{e}{2}(v_1{}^2 - v_2{}^2))^2 + m^2(v_1{}^2 + v_2{}^2). \qquad (6.29)$$

As already discussed earlier in section 6.3, for $m^2 > e\kappa/2$, we shall have a minimum of the right hand side of equation (6.29) when $v_1 = v_2 = 0$. Hence, by equation (6.24) U equals identity, and we get $|vac'> = |vac>$ and, thus perturbative vacuum is stable for a substitution as in equation (6.28b). As already seen, we have

$$<vac'|V|vac'> = \kappa^2/2 \neq 0. \qquad (6.30)$$

Hence supersymmetry is broken for this vacuum, but gauge symmetry remains unbroken.

On the other hand, when $m^2 < e\kappa/2$, the right hand side of equation (6.29) has a minimum when $v_1 = 0$ and

$$v_2 = v = (2/e)(e\kappa/2 - m^2)^{1/2} \neq 0. \qquad (6.31)$$

Then we obtain that

$$<vac'|V|vac'>_{\min} = (2m^2/e^2)(e\kappa - m^2), \qquad (6.32)$$

which is positive, and lower than $\kappa^2/2$. One can easily see that conventional $\mathcal{L}_{\text{eff}}$ after vacuum destabilisation is in fact given by

$$\mathcal{L}_{\text{eff}} = U^{-1}\mathcal{L}U, \qquad (6.33)$$

where the expression for (6.33) becomes the same as quoted in equation (6.12a) along with the other terms omitted there. Hence as stated earlier, here both gauge symmetry and supersymmetry get broken, with the same mixings as earlier.

We thus see that phase transition in quantum field theory merely indicates the destabilisation of the vacuum configuration with the new vacuum, $|vac'>$, being related to the old vacuum, $|vac>$, through equation (6.20). U is formally unitary since $B^\dagger - B$ is antihermitian, and may some times need a regularisation since integration over infinite space volume is involved. This understanding of phase transition has physical implications with or without supersymmetry as discussed in Ref.[9]. It can also be extended to condensates with a coherent state type of construction for the same [10]. The method is nonperturbative since only equal time algebra is used, but is variational through trial functions. We shall again illustrate the same in Chapter 13 for temperature dependance of the results with condensates.

6.7 Grand Unification Models

We shall here consider some braod features of gauge symmetry as well as symmetry breaking in the context of grand unification models. We first note that the objective here is to include strong, electromagnetic and weak interactions in a single framework. Since Salam Weinberg theory for electroweak unification is at present completely in agreement with experimental results, we shall expect that any grand unification symmetry group G we consider should contain the Salam Weinberg group as a subgroup, and for this group G the distinctive features of strong and electroweak interactions at a phenomenological as well as conceptual level should disppear. The simple groups which have been tried as examples for grand unification group G are the special unitary group $SU(5)$, the special orthogonal group $SO(10)$, the exceptional group E_6 as well as some other semisimple as well as simple groups. In the above, special means that the determinant of the corresponding matrices is unity. An excellent account of the description of these groups as well as the considerations of the same in the context of symmetry breaking have been given in the review article by Slansky [11]. We shall here give just a flavor of what can occur before dealing with some of these things in any detail later.

6.7.1 $SU(5)$

Grand unification theory with $SU(5)$ was originally proposed by Georgi and Glashow [12] and a supersymmetric generalisation of the same has been just discussed in subsection 6.4. Here the 24-plet adjoint representation yields the symmetry breaking to the standard model. This group is the minimal grand unification group, but is at present in contradiction with observations since it predicts the decay rate of proton going to e^+ and π^0 with a life-time much larger than the limits set by very careful experiments. The less sophisticated supersymmetric versions of the model also have the same defect.

6.7.2 $SO(10)$

The $SO(10)$ model is a grand unification model which has a very rich structure for intermediate symmetries. This group contains e.g. as subgroups the model with quark lepton unification with lepton number being the fourth color as was proposed for unification by Pati and Salam [13], some general left right symmetric models, as well as the minimal $SU(5)$ model as possible intermediate symmetries. In the context of supergravity we shall see in Chapter 10 the richness of the models associated with this group. The intermediate symmetry groups as well as the symmetry breaking patterns are too varied for us to single out any one of them here with more detail. The matter representation here corresponds to the 16-plet spinor representation of the group and thus as opposed to $SU(5)$ contains an extra right handed neutrino associated with symmetry breaking at an intermediate scale.

6.7.3 E_6 and E_8

The exceptional group E_6 with 78 generators was initially thought to be an extremely large group for unification. However, it became relevant as a subgroup of the exceptional group of rank eight, i.e. E_8 since such a symmetry appeared to get generated for superstring based theories. Although they are still favored on many considerations, the fact that they are very large groups makes them unmanageable to obtain even qualitative predictions. For the E_6 symmetry group the matter sector consists of a 27-plet representation, at least eight of which shall reflect the presence of mass scales not yet observed. Here we regard the fifteen quarks and leptons and the four particles corresponding to the two Higgs weak doublets as already "observed".

For the E_8 symmetry group, both matter and gauge sectors should belong to a supersymmetric 248-plet representation, which is a unique feature for a gauge group. This symmetry group has a very general mathematical structure, but we are completely at sea to find a path by which we can come from this structure (in ten dimensions) to the electroweak group in four dimensions with any degree of predictability.

We do not lack here ideas; in fact we have too many of them with no way of recognising which of them have a greater chance of being correct than the others. We shall discuss some of these in later chapters.

6.7.4 Proton decay, R-symmetry and R-parity

In any grand unification theory as mentioned above, there will be generally many particles with masses of the order of 10^{12} GeV or even of the Planck scale of 10^{19} GeV. Parallel to the case of weak interactions being caused by elimination of the $W^{\pm}$ mesons giving rise to a four fermion interaction, the elimination of the above fields shall give rise to unusual type of effective interactions at lower energies, some of which shall make proton unstable [14]. Supersymmetry presents here a fresh dimension to this instability of the proton even when we confine our attention to the particle multiplets of electroweak symmetry, while remembering that elimination of the heavy modes shall give rise to effective higher dimension operators, just as the elimination of the weak vector mesons gives rise to the dimension six effective Fermi interaction. Since supersymmetry implies the presence of spin zero partners for the quarks and leptons, we may have e.g. a dimension five operator with two quarks going over to the scalar partners of quarks or leptons. A dimension five operator as parametrised by a heavy mass scale will give rise to a much larger life time than a dimension six operator, since the heavy mass dimension will occur in the denominator in the coefficients of these higher dimension operators arising e.g. from the propagators for the heavy masses. Supersymmetry will give rise to many such effective interactions with fermion external lines replaced by scalar external lines yielding effective interactions which will make the proton decay much faster than what is observed. In order to avoid the presence of such interaction terms surviving at the low scales, a global symmetry called R-symmetry has been postulated.

A transformation corresponding to R-symmetry consists of a global $U(1)$ transformation $exp(-i\alpha)$ for the supercoordinates θ along with a similar transformation for the chiral superfield. Thus the chiral superfield Φ is said to have R-charge n when for a transformation as above,

$$R\Phi(x,\theta) \to e^{2in\alpha}\Phi(x,e^{-i\alpha}\theta), \qquad (6.34a)$$

and

$$R\Phi^{\dagger}(x,\bar{\theta}) \to e^{-2in\alpha}\Phi^{\dagger}(x,e^{i\alpha}\bar{\theta}). \qquad (6.34b)$$

A remark regarding R-symetry in the context of supersymmetric quantum mechanics in Chapter 2 may be worthwhile. There the action integral (2.11) is invariant under the transformation of the supercoordinate F given as

6.7. Grand Unification Models

$$F(t,\theta,\bar{\theta}) \to F(t, e^{-i\alpha}\theta, e^{i\alpha}\bar{\theta}). \qquad (6.35)$$

We may thus recognise that the theory is invariant under R-symmetry with the supercoordinate having R-charge zero. The generator of this transformation in fact is $\bar{\psi}\psi$, and thus a reflection of this symmetry was that $(d/dt)(\bar{\psi}\psi) = 0$. This was utilised in equation (2.18b) for quantisation of the fermionic coordinates.

We postulate that the Lagrangian shall be invariant under R-symmetry. Then, by assigning suitable R-charge to some chiral multiplets, some interactions which would be otherwise permissible and will give rise to rapid proton decay can now be forbidden. R-symmetry is expected to remain unbroken, so that we do not have a zero mass scalar Goldstone particle associated with it. The superpotentials giving rise to the F-type terms of the Lagrangian shall have R-charge two. The total R-charge of the action integral along with a $d^2\theta$ for the F-type terms has to get appropriately neutralised and thus the R-charge of superpotential must neutralise the R-charge of $d^2\theta$.

R-parity is a much simpler concept than R-symmetry, and is much less restrictive. We define a multiplicative quantum number called R-parity which is even for all the old nonsupersymmetric particles, but is odd for the new particles introduced by supersymmetry. Thus. e.g. quarks and leptons, the scalar Higgs particles, the weak vector mesons, have all even R-parity, whereas their superpartners as given in Table I of last chapter have odd R-parity. It is assumed that R-parity is conserved. Then obviously the superparticles will be produced in pairs. This feature quite often is adequate to avoid interaction terms yielding rapid proton decay. In addition, this has an extremely relevant experimental conclusion - the lightest supersymmetric particle will be absolutely stable, since it has odd R-parity and there is no other lighter particle with odd R-parity! R-parity is a main consideration in our attempts to discover supersymmetric particles; it is however unfortunate that till now we have no available experimental signature of any thing associated with the same.

We shall come back to the consideration of grand unification theories in Chapter 10 in the context of supergravity in four dimensions and in Chapter 12 in the context of supergravity in higher dimensions.

REFERENCES

1. L. O'Raifeartaigh, Nucl. Phys. B96, 331 (1975).

2. P. Fayet and J. Iliopoulos, Phys. Lett. B51, 461 (1974).

3. S. Dimopoulos and H. Georgi, Nucl. Phys. B193, 150 (1981).

4. N. Sakai, Z. Phys. C11, 153 (1981).

5. E. Witten, Phys. Lett. 105B, 267 (1981).

6. L. Girardello and M.T. Grisaru, Nucl. Phys. B194, 65 (1982).

7. R. Barbieri, S. Ferrara and C.A. Savoy, Phys. Lett. 119B, 343 (1982).

8. H.P. Nilles, Phys. Rep. 110, 1 (1984).

9. S.P. Misra, in *Proceedings of the workshop on high energy physics phenomenology*, edited by D.P. Roy and P. Roy, World Scientific, Singapore, 1989, p.346.

10. H. Mishra, S.P. Misra and A. Mishra, Int. J. Mod. Phys. A3, 2311 (1988); A. Mishra, H. Mishra, S.P. Misra and S.N. Nayak, Phys. Rev. D44, 110 (1991); ibid, Pramana (J. of Phys) 37, 59 (1991).

11. R. Slansky, Phys. Rep. 79, 1 (1981).

12. H. Georgi and S.L. Glashow, Phys. Rev. Lett. 32, 438 (1974).

13. J.C. Pati and A. Salam, Phys. Rev. D8, 1240 (1873); D10, 275 (1974).

14. J.C. Pati and A. Salam, Phys. Rev. Lett. 31, 661 (1973); S. Weinberg, Phys. Rev. D26, 287 (1982).

Chapter 7

Functional Methods in Superspace

7.1 Integration of Grassmann Variables

It is noted that path integral or functional methods in field theory quite often yield results which are nonperturbative, and hence generally prove to be more useful than perturbative methods. We wish to discuss here such methods. For this purpose, it is necessary to define integration over superspace, and hence, of Grassmann variables [1].

Normally integration is defined through a definition of measure in a certain space. However, Grassmann variables are of an entirely different type, and, as seen earlier, so are the functions of Grassmann variables. For example, if η is a Grassmann Variable, then the most general function of η is $f(\eta) = a + b\eta$. Hence, integration over Grassmann variables are to be specially defined so that they become post facto useful.

For a single Grassmann variable η we have already defined the following basic integrals in Chapter 2, [1]

$$\int d\eta = 0, \int \eta \, d\eta = 1. \qquad (7.1)$$

Hence, if we have $f(\eta) = a + b\eta$, then we obtain that

$$\int f(\eta) \, d\eta = \int (a + b\eta) \, d\eta = b. \qquad (7.2)$$

Dirac δ-function for Grassmann variables, $\delta(\eta)$ has to be so defined that

$$\int f(\eta) \, \delta(\eta) \, d\eta = f(0) = a.$$

Hence it is obvious that

$$\delta(\eta) = \eta. \qquad (7.3)$$

We shall define the integration elements for (θ^1, θ^2) space and $(\bar{\theta}_{\dot{1}}, \bar{\theta}_{\dot{2}})$ space through

$$d^2\theta = -\frac{1}{4}d\theta^\alpha\, d\theta^\beta\, \epsilon_{\alpha\beta} \qquad (7.4\text{a})$$

and

$$d^2\bar{\theta} = -\frac{1}{4}d\bar{\theta}_{\dot{\alpha}}\, d\bar{\theta}_{\dot{\beta}}\, \epsilon^{\dot{\alpha}\dot{\beta}}. \qquad (7.4\text{b})$$

We then easily see that

$$\int (\theta\theta)\, d^2\theta = \int (\bar{\theta}\bar{\theta})\, d^2\bar{\theta} = 1. \qquad (7.5)$$

We also substitute

$$d^4\theta = d^2\theta\, d^2\bar{\theta}. \qquad (7.6)$$

Using superspace integrals, we thus note that we can write for the chiral superfield parallel to equation (4.30a,b),

$$\mathcal{L} = \int \Phi^\dagger \Phi\, d^4\theta + \left[\int g(\Phi)d^2\theta + h.c.\right], \qquad (7.7)$$

where $g(\Phi)$ is the superpotential. Clearly, the D-type and F-type terms are respectively obtained by integrating over $d^4\theta$ and $d^2\theta$ respectively for appropriate superfields.

7.2 Superspace Dynamics

The procedure here will be the conventional method of path integrals. Thus, we shall first obtain the action integral in superspace, and then write the partition function in terms of the sources. The derivatives of the partition function will yield the Greens function in superspace, which in turn shall be used to evaluate the propagators in superspace and thus the contributions to Feynman diagrams.

Let us consider, as in Chapter 5, a chiral multiplet Φ and a vector multiplet $V = V^a t_a$ with t_a's as the generators of some gauge group G. Then, clearly the kinetic term of the chiral field along with the gauge interaction yields the action integral as the D-type contribution

$$\int d^4x\, d^4\theta\, \Phi^\dagger e^{\tilde{g}V} \Phi, \qquad (7.8)$$

where $\tilde{g}$ is the gauge coupling constant. In the following for the sake of simplicity *we shall omit the consideration of vector superfields* and take in equation (7.8) the gauge coupling to be zero. This can be included but will make the algebra more cumbersome.

If we have a superpotential $g(\Phi)$, then the F-type contribution to action is given as

7.2. Superspace Dynamics

$$\int d^4x \, d^2\theta \, g(\Phi). \tag{7.9}$$

We would however prefer to use all along $d^4\theta$ integrations. We can do this by introducing in the above equation $\delta(\bar{\theta})$. However, since we wish to maintain symmetry, we shall instead use the result for a *chiral f* such that

$$\int d^4x \, d^2\theta \, f = \int d^8z \, \frac{D^\alpha D_\alpha}{4\Box} \, f, \tag{7.10}$$

where $z = (x, \theta, \bar{\theta})$ and $\Box = \eta^{mn} \partial_m \partial_n$. In that case all $(\theta\theta)$ or $d^2\theta$ integration can be written in the same manner as the D-type terms. In fact, in order to obtain (7.10), let us first note that for any superfield f,

$$\int d^4x \, d^2\theta \, (-\bar{D}^2/4) \, f = \int d^8z \, f. \tag{7.11}$$

subject to neglecting surface terms for space-time integration. We may also note that for a *chiral* superfield f,

$$D^2 \bar{D}^2 f = -16\Box f, \tag{7.12}$$

where we have used the algebra of $\bar{D}_{\dot{\alpha}}$ and D_α as from equations (4.8). Equations (7.11) and (7.12) yield (7.10) in a formal manner. Using the above, for chiral fields we obtain the free field action [2,3,4]

$$S_0 = \int d^8z \left\{ \frac{1}{2}[\Phi, \Phi^\dagger] \begin{bmatrix} -(mD^2/4\Box) & 1 \\ 1 & -(m\bar{D}^2/4\Box) \end{bmatrix} \begin{bmatrix} \Phi \\ \Phi^\dagger \end{bmatrix} \right. \\ \left. + [\Phi, \Phi^\dagger] \begin{bmatrix} (D^2/4\Box)J \\ (\bar{D}^2/4\Box)J^\dagger \end{bmatrix} \right\}. \tag{7.13}$$

Clearly in the above we have included the mass terms which were F-type through the above identities. We now use the notations

$$\Psi = \begin{bmatrix} \Phi \\ \Phi^\dagger \end{bmatrix} \tag{7.14a}$$

$$B = \begin{bmatrix} (D^2/4\Box)J \\ (\bar{D}^2/4\Box)J^\dagger \end{bmatrix}. \tag{7.14b}$$

and

$$A = \begin{bmatrix} -(mD^2/4\Box) & 1 \\ 1 & -(m\bar{D}^2/4\Box) \end{bmatrix}. \tag{7.14c}$$

We then obtain the Euclidean partition function for the free chiral superfield as

$$Z_0(J.J^\dagger) = N\int D\Phi D\Phi^\dagger exp\left[\int(\frac{1}{2}\Psi^T A\Psi + \Psi^\dagger B)d^8z\right]$$
$$= N'exp\left[-\frac{1}{2}B^T A^{-1}B\, d^8z\right], \tag{7.15}$$

where

$$A^{-1} =$$

$$\begin{bmatrix} (m\bar{D}^2/4(\Box+m^2)) & 1+m^2\bar{D}^2 D^2/(16\Box(\Box+m^2)) \\ 1+m^2 D^2\bar{D}^2/(16\Box(\Box+m^2)) & (mD^2/4(\Box+m^2)) \end{bmatrix}.$$

$$\tag{7.16}$$

To derive the above, we may note the identities

$$D^2\bar{D}^2 D^2 = -16\Box D^2 \tag{7.17a}$$

and

$$\bar{D}^2 D^2 \bar{D}^2 = -16\Box\bar{D}^2. \tag{7.17b}$$

Using the above identities we thus obtain

$$\left[1 - m^2 D^2\bar{D}^2/(16\Box^2)\right]^{-1}$$
$$= 1 - (m^2 D^2\bar{D}^2/16\Box^2)\left[1 - m^2/\Box + \cdots\right]$$
$$= 1 + (m^2 D^2\bar{D}^2/16\Box(\Box+m^2)). \tag{7.18}$$

From equations (7.14b), (7.15) and (7.16), we then obtain the explicit form of Z_0 as

$$Z_0 = N'exp\left[\int d^8z\left\{J^\dagger(\Box+m^2)^{-1}J + J(mD^2/(8\Box(\Box+m^2)))J\right.\right.$$
$$\left.\left. + J^\dagger(m\bar{D}^2/(8\Box(\Box+m^2)))J^\dagger\right\}\right]. \tag{7.19}$$

The above expression coupled with the functional differentiation rule for the chiral sources as

$$\frac{\delta}{\delta J(z_1)}J(z_2) = -(\bar{D}_1^2/4)\delta_8(z_1-z_2), \tag{7.20a}$$

7.2. Superspace Dynamics

then yields the propagators for the superfield pairs $\Phi^\dagger(z_1)$ and $\Phi(z_2)$, $\Phi(z_1)$ and $\Phi(z_2)$, and, $\Phi^\dagger(z_1)$ and $\Phi^\dagger(z_2)$. Equation (7.20) is easily derived when we note that the chiral fields are functions of y and θ as defined in chapter 4, and thus the left hand side of equation (7.20) is

$$\delta_4(y_1 - y_2)\delta_2(\theta_1 - \theta_2)$$
$$= \delta_4(x_1 - x_2)(-\bar{D}^2/4)\delta_2(\bar{\theta}_1 - \bar{\theta}_2)\delta_2(\theta_1 - \theta_2). \quad (7.20b)$$

We may note that in the above θ_i and $\bar{\theta}_i$ are each Weyl spinors, and i stands for the vertices and not for the components.

From equation (7.19), by differentiating with respect to the chiral sources, we can identify the "free field" propagators in superspace as [2], without including the effect of equation (7.20),

$$< vac|T(\Phi(z_1)\Phi^\dagger(z_2))|vac > = -i(\Box + m^2)^{-1}\delta_8(z_1 - z_2), \quad (7.21a)$$

and

$$< vac|T(\Phi(z_1)\Phi(z_2))|vac >$$
$$= (-i/4)mD_1^{\,2}(\Box(\Box + m^2))^{-1}\delta_8(z_1 - z_2). \quad (7.21b)$$

We note that the chiral superfield contains the scalar field ϕ, the spin 1/2 field χ, and the auxiliary field h which for free fields is given by $h = -m\phi^*$. The above superpropagators in equation (7.21a) contain information about all the propagators for the component fields as above. For this purpose, however we now include the extra factors like $(-\bar{D}^2/4)$ and $(-D^2/4)$ as in equation (7.20) respectively for the functional differentiation of chiral fields or antichiral fields at each vertex. When we include these factors, we in fact identify the above propagators as [5]

$$< vac|T(\phi(x_1)\phi(x_2)^\dagger)|vac >$$
$$\cong (-i/16)(\Box + m^2)^{-1}\bar{D}_1^{\,2}D_1^{\,2}\delta_8(z_1 - z_2)$$
$$= -i(\Box + m^2)^{-1}\delta_4(x_1 - x_2) \equiv \Delta_c(x_1 - x_2) \quad \text{say}, \quad (7.22a)$$

$$< vac|T(\chi_\alpha(x_1)\bar{\chi}_{\dot\alpha}(x_2))|vac >$$
$$\cong \frac{1}{2}(-i/16)D_{1\alpha}\bar{D}_1^{\,2}\bar{D}_{2\dot\alpha}D_2^{\,2}(\Box + m^2)^{-1}\delta_8(z_1 - z_2)$$
$$= -i\sigma^m{}_{\alpha\dot\alpha}(\partial/\partial x_1^m)\Delta_c(x_1 - x_2), \quad (7.22b)$$

and

$$< vac|T(h(x_1)h(x_2)^\dagger)|vac >$$
$$\cong (1/16)(-i/16)D_1^{\,2}\bar{D}_1^{\,2}D_2^{\,2}\bar{D}_2^{\,2}(\Box + m^2)^{-1}\delta_8(z_1 - z_2)$$
$$= \Box\Delta_c(x_1 - x_2). \quad (7.22c)$$

We note that in equations (7.22b) and (7.22c) we are using the equations

$$D_\alpha \Phi|_{\theta=\bar\theta=0} = \sqrt{2}\chi_\alpha, \qquad (7.22d)$$

and,

$$D^2 \Phi|_{\theta=\bar\theta=0} = -4h, \qquad (7.22e)$$

which has been used in writing the above results. Thus $\cong$ implies an equality after putting $\theta = \bar\theta = 0$. In deriving (7.22b), we have also used equations (4.8) and (7.17). We may further notice that there is a discrepancy between equation (7.22c) and what we shall obtain for the propagator of $h(x_1)$ with $h(x_2)^*$ if we use directly that $h = -m\phi^*$ as in equation (4.22). The difference between the two arises from the point like contribution from the terms hh^* and $m(\phi h + \phi^* h^*)$ respectively in equations (4.22a) and (4.22b) for the auxiliary field. The propagators for ϕh and $\phi^* h^*$ are nonzero [5] and can be derived from $\Phi(1)\Phi(2)$ and $\Phi^\dagger(1)\Phi^\dagger(2)$ superpropagators. For similar reasons, the mass term is absent in the numerator in equation (7.22b).

Initially for supersymmetric theories perturbative calculations were done after the auxiliary fields were eliminated. Thus for example from equation (4.23) the Yukawa coupling $-g(\chi\chi\phi + \bar\chi\bar\chi\phi^*)$ and the trilinear and the quartic couplings $-mg\phi\phi^*(\phi+\phi^*)$ and $-g^2(\phi^*\phi)^2$ contribute to the quadratic mass renormalisation contribution of mesons. When calculated with appropriate regularisation, the contributions when added cancel with each other. This looks "magical" since quite complicated expressions give this zero contribution. Wess and Zumino [5] modified the calculations retaining the auxiliary fields and then doing perturbation theory. This reflected the effect of supersymmetry more clearly in the corresponding perturbation theory. Superfield propagators as above go a step further in this direction.

In equation (7.19) we have written the partition function for the free fields. We shall now obtain the partition function for the interacting fields. The expression for the same is very similar to that of field theory [6]. If we have a superpotential $g(\Phi)$, in terms of equation (7.19) we obtain in fact the partition function for the interacting system as

$$Z(J, J^\dagger) = exp\left[-ig\left(-i\frac{\delta}{\delta J}\right) + ig\left(i\frac{\delta}{\delta J^\dagger}\right)\right] Z_0(J, J^\dagger). \qquad (7.23a)$$

The methodology for arriving at this expression is the same as in conventional field theory. We replace the free field action integral of equation (7.13) including the contribution to this action from $g(\Phi)$ and $g(\Phi)^\dagger$ and then expand the extra terms in the functional integral to convert the superfields to derivatives with respect to the sources. This gives the partition function Z in terms of Z_0 as written above. We then obtain

7.2. Superspace Dynamics

the corrections to the Greens functions in different orders of the Feynman diagrams by expanding the exponential in equation (7.23). We note that here in superspace the superpotential operates in a parallel manner as the interaction hamiltonian in conventional field theory. In equation (7.23) we have used the notation for example as

$$g\left(-i\frac{\delta}{\delta J}\right)Z_0(J,J^\dagger)$$
$$\equiv \int d^8z\, g\left(-i\frac{\delta}{\delta J(z)}\right)Z_0(J,J^\dagger), \qquad (7.23b)$$

which is to be substituted in the expansion of the exponential in equation (7.23). Thus each vertex gives rise to an integration in superspace corresponding to Feynman diagrams in superspace.

Let us consider the free superpropagators in the momentum space. We shall take the Fourier transforms only for the ordinary space dependant part, and shall leave the Grassmannian coordinates as they are. In that case we note that

$$D_\alpha(k,\theta,\bar\theta) = \frac{\partial}{\partial\theta^\alpha} - \sigma^m{}_{\alpha\dot\alpha}\bar\theta^{\dot\alpha}\, k_m, \qquad (7.24a)$$

$$\bar D_{\dot\alpha}(k,\theta,\bar\theta) = -\frac{\partial}{\partial\bar\theta^{\dot\alpha}} + \theta^\alpha\sigma^m{}_{\alpha\dot\alpha}\, k_m, \qquad (7.24b)$$

From the above we can show that while acting on $\delta_8(z_1-z_2)$,

$$D_{1\alpha} \equiv -D_{2\alpha}, \qquad (7.25)$$

where we have included the fact that in D_2 the sign of the four momentum k as in equation (7.24) is to be reversed since the flow of four momentum from 1 to 2 for a propagator is opposite of that from vertex 2 to vertex 1.

We now write the Feynman rules in momentum space for chiral multiplets in perturbation theory as below.

(i) For massive chiral superfields (scalar and ghost multiplets) the $\Phi\Phi^\dagger$ and $\Phi\Phi$ propagators are respectively given from equations (7.21) as

$$(p^2+m^2)^{-1}\delta_4(\theta_{12})$$

and

$$m\,(p^2)^{-1}(p^2+m^2)^{-1}\left(\frac{1}{4}\right)D^2(p,\theta_1,\bar\theta_1);$$

(ii) Vertices are read from the superpotential $g(\Phi)$ and have extra factors $(-\bar{D}^2/4)$ and $(-D^2/4)$ for each chiral or antichiral superfield arising from equation (7.20), with one factor $-\bar{D}^2/4$ or $-D^2/4$ omitted for converting the superspace integrations in $d^2\theta$ or $d^2\bar{\theta}$ to $d^4\theta$;

(iii) There are usual combinatorial factors, and -1 for each ghost loop since it is fermionic, which we have not considered;

(iv) For each external line one factor of $-\bar{D}^2/4$ or $-D^2/4$ shall be omitted for chiral or antichiral superfields respectively since here functional differentiation with respect to the sources is absent;

(v) There will be usual momentum space factors corresponding to the vertices and propagators with an appropriate dimensional regularisation.

As mentioned in the above statements we have not included the gauge multiplets, although we have stated about ghost chiral fields. The extension of the above discussions of path integral method for quantisation in superspace to gauge fields is conceptually similar to what we have done, but the applications as well as techniques are more complicated. We have, however, followed the format of Ref. [2] which also includes this in detail along with examples.

7.3 Cancellation of Divergences

We shall here calculate some higher order corrections to the propagators for the chiral superfields, which will demonstrate the power of the superspace formulation. We shall show that certain type of terms do not arise when we use this method. Any other method of calculation therefore shall give them as zero, and shall naturally lead to a cancellation of the individual contributions which otherwise might look "magical". Before we do this, however, we shall prove the nonrenormalisation theorem on which such results are based.

Theorem: *Each term in the effective action can be expressed as an integral with a single $d^4\theta$. Equivalently, the effective action is local in the superspace coordinates θ and $\bar{\theta}$.*

To prove the above, let us consider any loop in a one particle irreducible diagram with n vertices. Since we wish to show locality in superspace coordinates, we shall suppress the momentum variables which shall be as noted in the Feynman rules above. Let us use the notation that θ_i stands for $(\theta_i, \bar{\theta}_i)$ and $\theta_{ij} = (\theta_i - \theta_j, \bar{\theta}_i - \bar{\theta}_j)$. In superspace, we shall have the product of delta functions $\delta_4(\theta_{12})\delta_4(\theta_{23})\cdots\delta_4(\theta_{n1})$ on which the covariant derivatives D and $\bar{D}$ shall operate. We can have by integration in parts the operators D_1 and D_2 removed from the action

7.3. Cancellation of Divergences

on $\delta_4(\theta_{12})$ and make the same operate on $\delta_4(\theta_{n1})$ or on $\delta_4(\theta_{23})$ or on the external superfields of the loop as may be relevant for the specific contribution. We then perform the integration over θ_2 using $\delta_4(\theta_{12})$ and thus have $\delta_4(\theta_{23})$ now replaced by $\delta_4(\theta_{13})$. We can next repeat the same process for θ_3 integration, and, proceeding further, thus complete integrations over the whole set $\theta_2, \theta_3, \cdots, \theta_{n-1}$. We shall then have sum of expressions like

$$\int d^4\theta_1 d^4\theta_n \delta(\theta_1) f(\theta_1, \theta_n) [D \cdots \bar{D}] \delta_4(\theta_{n1})$$
$$= \int d^4\theta_1 f(\theta_1, \theta_n) [D \cdots \bar{D}] \delta_4(\theta_{n1})\big|_{\theta_n = \theta_1} \qquad (7.26a)$$

The evaluation of the above expression and integration over θ_1 is again straightforward and relatively simple when we use the anticommutivity properties of the cavariant derivatives. Thus for the above loop we are left with a single integration in the space of supercoordinates. When there are many loops we can repeat the process loop by loop. This completes the proof of the above theorem.

While using this theorem we can easily see that this leads to many types of contributions to vanish resulting from the properties of the covariant derivatives D and $\bar{D}$. This shows that some contributions in supersymmetric theories are zero leading to nonrenormalisation for the corresponding processes. This justifies the name of the theorem.

We can further see that in coordinate space the general form of the effective action becomes

$$\int d^4x_1 \cdots d^4x_n d^4\theta F_1(x_1, \theta) \cdots F_n(x_n, \theta). \qquad (7.26b)$$

From the fact that the F's are constructed only from the covariant derivatives, it can be seen that the above action is invariant under the supersymmetry transformations as defined in Chapter 4 in equation (4.1). Thus the contributions are manifestly supersymmetric. This was the reason for choosing supersymmetric covariant derivatives in the construction of the effective action. The method e.g. can not be implemented with gauge covariant derivatives in ordinary space for the construction of the S-matrix for the interacting system except for some rare cases where only a formal solution can be obtained using functional mehtods. The trick was here useful and practicable since the variables were Grassmannian, and thus integrations and differentiations are relatively simple as well as self-limiting.

Let us now consider a trivial application of the above technology. Let us take the superpotential as

$$g(\Phi) = \frac{\lambda}{3!} \Phi^3. \qquad (7.27)$$

considered earlier in equation (4.20c) in Chapter 4. We shall now calculate the lowest order correction to the $\Phi(1)\Phi^\dagger(2)$ propagator for massless chiral superfields. In superspace this will involve two external lines giving rise to superfields $\Phi^\dagger(-p,\theta_1)$ and $\Phi(p,\theta_2)$ along with two $\Phi^\dagger\Phi$ propagators with integration over a four momentum k for the loop momentum. Normally the antichiral field vertex 1 shall contain the operator $-D^2/4$ thrice explicitly corresponding to the expression (7.27). But here we have one external antichiral field, and hence one less $-D^2/4$. Further, another $-D^2/4$ gets absorbed in converting the $d^2\bar{\theta}$ to $d^4\theta$. A similar comment also holds for the chiral superfields at vertex 2. In fact, using the Feynman rules as above, the expression for the corresponding contribution then becomes

$$\Gamma(\Phi,\Phi^\dagger) = \frac{1}{2}\lambda^2 \frac{1}{(2\pi)^8} \int d^4p\, d^4k\, d^4\theta_1\, d^4\theta_2$$
$$\Phi^\dagger(-p,\theta_1)\,\Phi(p,\theta_2)\frac{1}{[k^2]}\frac{1}{[(k+p)^2]}$$
$$\delta_4(\theta_{12})(1/16)\bar{D}^2 D^2(k,\theta_2)\delta_4(\theta_{12}). \tag{7.28}$$

As stated in the theorem above, we have made one δ-function in supercoordinates free, so that it can be integrated out and the above expression evaluated. To be more exact, we should take $d^D k/(2\pi)^D$ for dimensional regularisation, and then finally put $D=4$ for four dimensions instead of what is written in equation (7.28). When we do so, and substitute

$$\bar{D}^2 D^2 \delta_4(\theta_{12})\big|_{\theta_1=\theta_2=0} = 16,$$

we get that

$$\Gamma(\Phi,\Phi^\dagger) = \frac{1}{2}\frac{\lambda^2}{(2\pi)^4} \int d^4p\, d^4\theta\, A(p)\, \Phi^\dagger(-p,\theta)\, \Phi(p,\theta), \tag{7.29}$$

where

$$A(p) = \frac{1}{(2\pi)^D}\int \frac{d^D k}{[k^2][(k+p)^2]}$$
$$= \frac{1}{(4\pi)^{D/2}}\Gamma\!\left(2-\frac{D}{2}\right)\left[\Gamma\!\left(\frac{D}{2}-1\right)\right]^2 \frac{1}{\Gamma(D-2)}(p^2)^{(D/2)-1}. \tag{7.30}$$

Clearly the pole of $\Gamma(2-(D/2))$ for $D=4$ implies the known divergence contained in this expression. This divergence is eliminated through a wave function renormalisation [6]. It was calculated by Wess and Zumino [5] earlier using the auxiliary fields but not using the superspace formulation.

7.3. Cancellation of Divergences

Let us next calculate the lowest order contribution with two chiral superfields as external lines. In that case we shall have two $\Phi\Phi$ super propagators which do not vanish and have been derived earlier in equation (7.21b). However, parallel to the last line on the right hand side of equation (7.28), there will be a contribution with $(D^2)^2$ instead of $\bar{D}^2 D^2$. This will give zero while acting on $\delta_4(\theta_{12})$. Thus the correction term for the effective action will vanish. From equation (4.20b) we recognise that such a correction would have given rise to mass renormalisation which now vanishes. By the nonrenormalisation theorem the same can also be seen to be true when higher order corrections are retained. This is an example of absence of renormalisation in supersymmetry and here gives rise to the stability of mass hierarchy against renormalisation effects. This feature was one of the primary motivations for introducing supersymmetry for grand unification theories which contain disparate mass scales.

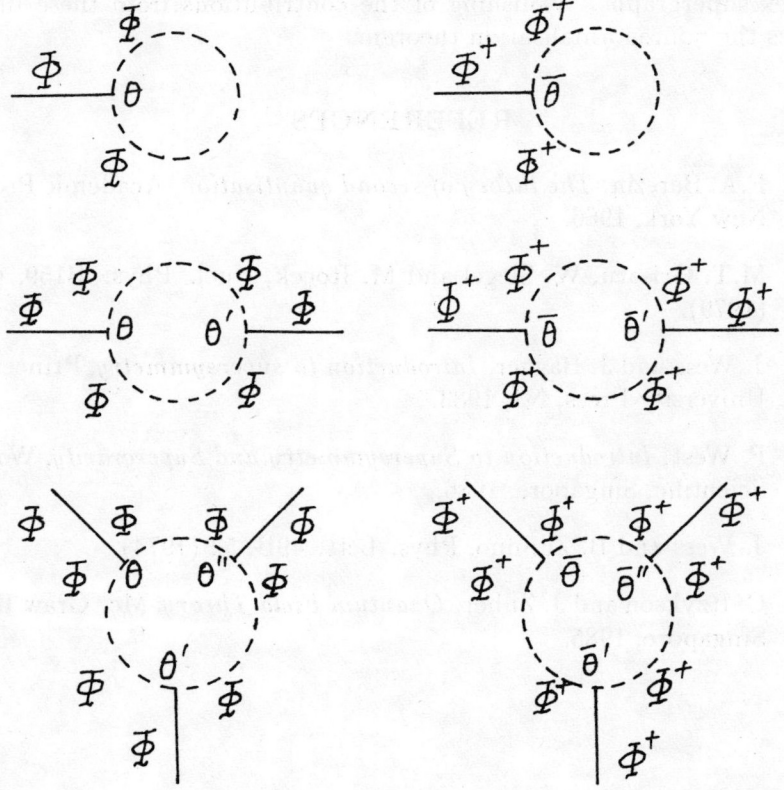

Fig. 1

In the above we have ignored the presence of Yang Mills fields through vector superfields introduced in Chapter 5. The introduction of the same is conceptually straightforward, but makes the calculations more cumbersome. There we have to obtain a convenient structure for the super propagators for the vector super fields, and then introduce the interaction with chiral (matter) superfields through the expansion of the exponential in equation (5.25) or (7.8), using that this expansion terminates in Wess Zumino gauge. We have also to introduce the ghost superfields. The statements made in the course of our discussions above like the nonrenormalisation theorem, which only depends on the manifestly supersymmetric covariant formulation of the Feynman rules, continues to hold good. The consideration of the same may be found in Refs. [2,4]. Our objective here has been to keep the discussions at a sufficiently elementary level for the beginners and hence omit them here. We however give some simple graphs to illustrate the cancellation of divergences as in Fig. 1 for the lowest order tadpole, self-energy, and vertex supergraphs. Vanishing of the contributions from these illustrates the nonrenormalisation theorem.

REFERENCES

1. F.A. Berezin, *The theory of second quantisation*, Academic Press, New York, 1966.

2. M.T. Grisaru, W. Siegel and M. Rocek, Nucl. Phys. B159, 429 (1979).

3. J. Wess and J. Bagger, *Introduction to supersymmetry*, Princeton University Press, NJ, 1983.

4. P. West, *Introduction to Supersymmetry and Supergravity*, World Scientific, Singapore, 1986.

5. J. Wess and B. Zumino, Phys. Lett. 49B, 52 (1974).

6. C. Itzykson and J. Zuber, *Quantum Field Theory*, Mc. Graw Hill, Singapore, 1985.

Chapter 8

Superspace and Supergravity: A Geometric Picture

8.1 Introduction

We first note that supergravity is a specific way of considering local supersymmetry. Thus the "increments" ξ and $\bar{\xi}$ in equation (4.1) which are Grassmannian should now be functions of space time variable x. We had earlier seen that superspace techniques are conceptually and otherwise richer to describe supersymmetry. Hence we shall adopt those techniques for the definition of local supersymmetry, i.e. supergravity. For this purpose, the language of differential geometry is quite useful, which we shall now proceed to discuss.

8.2 Differential forms in Superspace

We shall consider a space of eight variables [1] with coordinates $z^M = (x^m, \theta^\mu, \bar{\theta}_{\dot{\mu}})$. In the above m runs through (0,1,2,3), the space-time indices, and μ through 1 and 2, the spinorial indices. We now consider the superspace (or the supermanifold) as the space of all z^M as above along with their differentials dz^M for the tangent space. We then define the multiplication rule for these objects as

$$z^M z^N = (-1)^{mn} z^N z^M, \qquad (8.1a)$$

$$dz^M dz^N = -(-1)^{mn} dz^N dz^M, \qquad (8.1b)$$

$$dz^M z^N = (-1)^{mn} z^N dz^M. \qquad (8.1c)$$

In the above $m = 0$ if M is a space-time index and $m = 1$ if M is a spinorial index. Equation (8.1a) in the above tells us that the space-time variables, which are ordinary numbers, commute with each other and with the Grassmannian variables, whereas the Grassmannian variables anticommute with each other, a property which we have used earlier. Equation (8.1b) means that the differentials of the space-time variables anticommute with each other, as in ordinary differential geometry, and so do the differentials of the Grassman variables and ordinary variables, whereas the differentials of the Grassman variables commute with each other. The meaning of equation (8.1c) can be similarly deciphered.

We next define a p-form Ω as

$$\Omega = dz^{M_1} \cdots dz^{M_p} f_{M_p \cdots M_1}(z). \tag{8.2}$$

Clearly the function $f_{M_p \cdots M_1}(z)$ will have symmetry properties as governed by equation (8.1b). The exterior derivative $d\Omega$ of the above form is defined as

$$d\Omega = dz^{M_1} \cdots dz^{M_p} dz^{M_{p+1}} \frac{\partial}{\partial z^{M_{p+1}}} f_{M_p \cdots M_1}(z). \tag{8.3}$$

Thus the exterior derivative of a p form is a $p+1$ form. If Σ is a q form, we may easily see that $\Omega\Sigma$ is a $p+q$ form. We may also verify that

$$d(\Omega\Sigma) = \Omega d\Sigma + (-1)^q (d\Omega)\Sigma. \tag{8.4}$$

We may also have a structure group associated with differential forms. Let G be a group and $X_b{}^a(z)$ be any matrix element corresponding to a specific representation of the group which also is a function of z. Let Ω^a be a p form transforming as

$$\Omega'^a = \Omega^b X_b{}^a, \tag{8.5a}$$

or symbolically

$$\Omega' = \Omega X. \tag{8.5b}$$

We next note that if X is an infinitesimal transformation, then we have

$$X_b{}^a = \delta_b{}^a + i(t_i)_b{}^a \epsilon_i, \tag{8.6}$$

where ϵ_i's are the real parameters describing the group transformations, and the matrices t_i's are the generators of the group for the specific representation with i running from 1 to k with k as the dimension of the group.

We shall now define the covariant derivative $D\Omega$ of the form Ω given as

$$D\Omega = d\Omega + \Omega\phi. \tag{8.7}$$

In the above, ϕ is a 1-form called connection defined as

8.2. Differential forms in Superspace

$$\phi_a{}^b = dz^M \phi_{Ma}{}^b. \tag{8.8}$$

We wish the covariant derivative to transform as

$$D'\Omega' = (D\Omega)X, \tag{8.9}$$

which decides the transformation property for the connection ϕ as

$$\phi' = X^{-1}\phi X - X^{-1}dX. \tag{8.10}$$

We may note that the gauge fields transform as above, and shall see later that they are in fact connections. We may also note that connections are *not tensors* since they transform differently as seen above.

We take the eight dimensional superspace as a manifold. Hence the space is locally isomorphic to an "Euclidian" space. This feature is incorporated through the "vielbein" 1-forms defined as

$$E^A = dz^M E_M{}^A(z). \tag{8.11}$$

We note that the indices $M, N, \cdots$ describe the coordinates as earlier in a given "coordinate patch", and the indices $A, B, \cdots$ describe, loosely speaking, the "local coordinates" of gravity. Thus we call the former Einstein indices or world indices, and the later as Lorentz indices or group indices. In four dimensional gravity M runs over $0, 1, 2, 3$ and similarly for the group indices, which here is the Lorentz group. $E_A{}^M$ is the inverse of vielbein taken earlier so that we have

$$E_A{}^M E_M{}^B = \delta_A{}^B, \quad E_M{}^A E_A{}^N = \delta_M{}^N.$$

We note in particular the $E_M{}^A$ is a *contravariant* Lorentz vector and a *covariant* Einstein vector. We shall freely use these vielbein to convert Lorentz indices to Einstein indices and vice versa. Thus e.g. we write

$$\phi_{MN}{}^L = \phi_{MA}{}^B E_N{}^A E_B{}^L, \tag{8.12}$$

where both the vielbein and its reciprocal have been used. In four dimensions under certain conditions we can identify that

$$\phi_{nl}{}^m = \Gamma_{nl}{}^m, \tag{8.13}$$

where Γ is the Christoffel symbol of the second kind. However, the connection in a manifold generally is not uniquely defined [2]. The curvature is defined as the 2-form

$$F_A{}^B = d\phi_A{}^B + \phi_A{}^C \phi_C{}^B$$
$$= (1/2)dz^M dz^N F_{NMA}{}^B, \tag{8.14}$$

where $F_{MNA}{}^B$ are the components of the curvature tensor. We may note that under a group transformation

$$F' = d\phi' + \phi\phi = X^{-1}FX, \qquad (8.15)$$

which shows that curvature is a tensor. From equation (8.14) we have explicitly

$$\begin{aligned} F_{NMA}{}^B &= \partial_N \phi_{MA}{}^B - (-1)^{mn} \partial_M \phi_{NA}{}^B \\ &\quad + (-1)^{n(m+a+c)} \phi_{MA}{}^C \phi_{NC}{}^B \\ &\quad - (-1)^{m(a+c)} \phi_{NA}{}^C \phi_{MC}{}^B. \end{aligned} \qquad (8.16)$$

In the above we have taken the local group to be the "Lorentz group" extended to the superspace. In four dimensional gravity the above equation reduces to

$$\begin{aligned} F_{nma}{}^b &= \partial_n \phi_{ma}{}^b - \partial_m \phi_{na}{}^b \\ &\quad + \phi_{ma}{}^c \phi_{nc}{}^b - \phi_{na}{}^c \phi_{mc}{}^b. \end{aligned} \qquad (8.17)$$

The above justifies the name "curvature". We may identify the same with the notations of Weinberg [3] through

$$R^B{}_{AMN}\big|_{\text{Weinberg}} = F_{NMA}{}^B. \qquad (8.18)$$

One more important concept here is that of the 2-form torsion defined as

$$\begin{aligned} T^A &= DE^A = dE^A + E^C \phi_C{}^A & (8.19a) \\ &= (1/2) E^C E^B T_{BC}{}^A & (8.19b) \\ &= (1/2) dz^N dz^M T_{MN}{}^A. & (8.19c) \end{aligned}$$

In ordinary gravity, torsion is usually taken to vanish. However, we shall see that in supergravity it plays a vital role.

We may note that curvature and torsion are related by Bianchi identities given as

$$E^B R_B{}^A = DT^A \qquad (8.20)$$

where both left hand side and right hand side are equal to DDE^A.

8.3 Local Supersymmetry

Parallel to equation (4.1) we define local supersymmetric transformations through

$$x^m \to x^m + i\,\bar{\theta}\gamma^m \xi(x) \qquad (8.21a)$$

and

8.3. Local Supersymmetry

$$\theta^\mu \to \theta^\mu + \xi^\mu(x). \tag{8.21b}$$

In the above we are using the *four component* notation for the spinors as described in the third chapter. The above is particular case of general coordinate transformation in superspace given as

$$z^M \to z^M + \xi^M(z). \tag{8.22}$$

This was the reason to get into the structure of general theory of relativity as envisaged in the last section. Before considering (8.22) let us consider global supersymmetry through the description of a flat eight dimensional supermanifold.

In flat superspace we identify the supersymmetric covariant derivative of equation (4.8) through the vielbein given as

$$D_A = e_A{}^M(z)\,\partial_M. \tag{8.23}$$

With the four component notation as described earlier, we then obtain for the spinorial components the nontrivial equation

$$D_\alpha = \frac{\partial}{\partial \theta^\alpha} + i\,(\gamma^m)_\alpha{}^\beta\,\theta_\beta\,\partial_m, \tag{8.24}$$

so that the vielbein for the flat eight dimensional superspace becomes [1,4]

$$e_A{}^M(z) = \begin{bmatrix} \delta_a{}^m & 0 \\ i(\gamma^m)_\alpha{}^\beta \theta_\beta & \delta_\alpha{}^\mu \end{bmatrix}, \tag{8.25a}$$

with the inverse given as

$$e_M{}^A(z) = \begin{bmatrix} \delta_m{}^a & 0 \\ -i(\gamma^a)_\mu{}^\nu \theta_\nu & \delta_\mu{}^\alpha \end{bmatrix}, \tag{8.25b}$$

As earlier we take $M = (m, \mu)$ and $A = (a, \alpha)$. We immediately notice that here *torsion does not vanish* and in fact is evaluated as

$$T_{\mu\nu}{}^a = i\,(\gamma^a)_{\mu\nu}. \tag{8.26}$$

We take here however the connection one form ϕ to be zero, so that curvature vanishes, and as mentioned, the space is flat.

We now go back to the general coordinate transformation (8.22) and write the vielbein superfield one form $E_M{}^A(z)$ as

$$E_M{}^A(z) = \begin{bmatrix} E_m{}^a(z) & \tfrac{1}{2}\psi_m{}^\alpha(z) \\ E_\mu{}^a(z) & E_\mu{}^\alpha(z) \end{bmatrix} \tag{8.27}$$

and also consider a connection superfield one form $\phi_{MA}{}^B(z)$. However we wish $\xi^M(z)$ to be restricted to equation (8.21). Also we want to have only the components $e_m{}^a(x)$ and $\psi_m{}^\alpha(x)$ as independent fields for the description of the gravity multiplet, with a minimal set of auxiliary fields. We must thus constrain the space of vielbein and connection in the eight dimensional super manifold in a consistent manner to obtain the above features. We shall here describe the broad geometric principles involved; the details may be found in Ref. [1].

We note that in $E_M{}^A(z)$, M is an Einstein tensor index and A is a Lorentz tensor index. Hence under an infinitesimal general coordinate transformation as in equation (8.22) and an infinitesimal Lorentz transformation L, we shall have,

$$\delta E_M{}^A = -\xi^L \, \partial_L E_M{}^A - (\partial_M \xi^L)\, E_L{}^A + E_M{}^B \, L_B{}^A \qquad (8.28)$$

which gives rise to

$$\delta_\xi E_M{}^A = -D_M \xi^A - \xi^B \, T_{MB}{}^A, \qquad (8.29)$$

where we have made the infinitesimal Lorentz transformation as ξ dependant, given as

$$L_B{}^A = -\xi^C \phi_{CB}{}^A. \qquad (8.30)$$

A general coordinate transformation followed by the field dependant Lorentz transformation (8.30) is known as a super gauge transformation [1].

We now wish to limit the form of the vielbein to

$$E_M{}^A(z)\big|_{\theta=0} = \begin{bmatrix} e_m{}^a(x) & \tfrac{1}{2}\psi_m{}^\alpha(x) \\ 0 & \delta_\mu{}^\alpha \end{bmatrix} \qquad (8.31)$$

along with a restriction on connection as

$$\phi_{mA}{}^B(z)\big|_{\theta=0} = \omega_{mA}{}^B(x) \qquad (8.32a)$$

and

$$\phi_{\mu A}{}^B(z)\big|_{\theta=0} = 0 \qquad (8.32b)$$

Consistent with global supersymmetry, parallel to (8.25) we shall also adopt the constraint

$$T_{\alpha\beta}{}^a = i\,(\gamma^a)_{\alpha\beta}. \qquad (8.33)$$

8.3. Local Supersymmetry

We then obtain the components of the connection interms of vielbein as

$$\begin{aligned}\phi_{nml}(z)|_{\theta=0} &= \frac{1}{2}\big[-\frac{i}{2}(\psi_n\,\gamma_m\,\psi_l - \psi_m\,\gamma_l\,\psi_n + \psi_l\,\gamma_n\,\psi_m)\\ &\quad - e_{la}\,(\partial_n e_m{}^a - \partial_m e_n{}^a) - e_{ma}\,(\partial_l e_n{}^a - \partial_n e_l{}^a)\\ &\quad + e_{na}\,(\partial_m e_l{}^a - \partial_l e_m{}^a)\big].\end{aligned} \qquad (8.34)$$

In the above the vielbein components on the right hand side coorespond to an old result of four dimensional gravity [3].

In the super manifold we may write the Ricci tensor R_{CA} as

$$R_{CA} = R_C{}^D{}_{AD} \qquad (8.35a)$$

and the scalar curvature $R(z)$ as

$$\begin{aligned}R &= g^{AC}\,R_{CA}\\ &= g^{ac}\,R_c{}^d{}_{ad} + C^{\alpha\gamma}\,R_\gamma{}^\delta{}_{\alpha\delta},\end{aligned} \qquad (8.35b)$$

with

$$C^{\alpha\beta} = \begin{bmatrix} \epsilon^{\alpha\beta} & 0 \\ 0 & \epsilon_{\dot\alpha\dot\beta} \end{bmatrix}. \qquad (8.36)$$

In the above we have used the symbols α, β both for *four components* on the left hand side as well as for *two components* as on the right hand side as earlier. We next note that from the Lorentz transformation properties we can identify the following structure for the spinorial objects here with two component notations as follows:

$$R_{\beta\delta\gamma\alpha} = \epsilon_{\gamma\beta}\,G_{\alpha\delta} + \epsilon_{\alpha\beta}\,G_{\gamma\delta}, \qquad (8.37)$$

where $G^*_{\alpha\dot\alpha} = G_{\alpha\dot\alpha}$, with the constraint here being written in terms of two component notation as compared to the four component notation for spinorial indices in equation (8.37). This now defines a real vector field with

$$G_{\alpha\dot\alpha}|_{\theta=0} = -(1/3)\,b_a\,\sigma^a{}_{\alpha\dot\alpha} \qquad (8.38)$$

as parallel to equation (3.17a). It has been examined in Ref. [1] that the above vector field $b_a(x)$ as well as the complex scalar field M defined through

$$R(z)|_{\theta=0} = -(1/6)\,M(x), \qquad (8.39)$$

can not be expressed in terms of the components of vielbein $e_m{}^a(x)$ or $\psi_m{}^\alpha(x)$.

8.4 Gravity multiplet

As noted, we shall have supergravity multiplet given as $e_m{}^a(x)$, $\psi_m{}^\alpha(x)$, $b_a(x)$, and $M(x)$. We now proceed to obtain the supergravity transformations. Clearly, as stated, this will consist of special form of general coordinate transformations $\xi^A(z)$, and, Lorentz transformations $L_A{}^B(z)$. We first restrict that

$$\xi^a(z)\big|_{\theta=0} = 0; \qquad (8.40\text{a})$$

$$\xi^\alpha(z)\big|_{\theta=0} = \xi(x); \qquad (8.40\text{b})$$

and, for the ξ dependant Lorentz transformation,

$$L_{AB}(z)\big|_{\theta=0} = 0. \qquad (8.40\text{c})$$

We next note that

$$\delta E_M{}^A(z) = -D_M \xi^A - \xi^B T_{MB}{}^A + E_M{}^B L_B{}^A. \qquad (8.41)$$

As explained in Ref.[1] in a straight forward manner, the vielbein can continue to have the restricted form of (8.31) provided we choose that

$$\xi^a(x) = 2i\theta\gamma^a\xi(x). \qquad (8.42)$$

This correlates a translation in space time with the spinorial supersymmetric transformation $\xi(x)$. Also, the connections can continue to satisfy the constraints of equations (8.32) provided we choose

$$L_{\alpha\beta}(x) = (1/3)\left\{\theta_\alpha\left(2\xi_\beta(x)M^*(x) - b_m\,\gamma^m{}_\beta{}^\gamma\,\xi_\gamma(x)\right)\right\} \qquad (8.43)$$

and similar expressions for the Lorentz transformations in other sectors consistent with the above [1]. We note that equation (8.43) is equivalent to two equations with the two component notations. Local supersymmetry transformations originate from the above general coordinate transformations with a local $\xi^\alpha(x)$, along with field dependent compensating Lorentz transformations as in (8.43). In this case, one obtains that the fields $e_m{}^a(x)$, $\psi_m{}^\alpha(x)$, $M(x)$ and $b_m(x)$ transform amongst each other, with the explicit forms being rather lengthy expressions.

8.5 Chiral and Vector Multiplets

We first note that for chiral fields two component notations are more convenient. As earlier, we now define chiral multiplets through the condition

8.5. Chiral and Vector Multiplets

$$\bar{D}_{\dot\alpha}\Phi = 0, \tag{8.44}$$

and next define component fields as

$$A = \Phi\big|_{\theta=0=\bar\theta} \tag{8.45a}$$

$$\chi_\alpha = (1/2) D_\alpha \Phi\big|_{\theta=0=\bar\theta} \tag{8.45b}$$

and

$$F = -(1/4)(DD)\Phi\big|_{\theta=0=\bar\theta}. \tag{8.45c}$$

We may note that we have to define spinors in curved space-time through Lorentz indices [2], as done above. With the supersymmetry transformation defined earlier, then we can show that [1]

$$\delta A = -2\xi^\alpha \chi_\alpha, \tag{8.46a}$$

$$\delta \chi_\alpha = -2\xi_\alpha F - 2i\sigma^a{}_{\alpha\dot\beta}\,\bar\xi^{\dot\beta} D_a A, \tag{8.46b}$$

and,

$$\delta F = -(2/3) M^* \xi^\alpha \chi_\alpha + \bar\xi^{\dot\alpha}\big((2/3) b_a\, \sigma^a{}_{\alpha\dot\alpha}\,\chi^\alpha - 2i\,\sigma^a{}_{\alpha\dot\alpha}\, D_a\,\chi^\alpha\big). \tag{8.46c}$$

In the above, the super covariant derivative D_a is defined through

$$D_a A = e_a{}^m \Big(\partial_m A - \frac{1}{2}\psi_m{}^\alpha \chi_\alpha \Big) \tag{8.47a}$$

and

$$D_a \chi_\alpha = e_a{}^m \Big(D_m \chi_\alpha - \frac{1}{2}\psi_{m\alpha} F - \frac{i}{2}\bar\psi_a^{\dot\beta}\,\sigma^b{}_{\alpha\dot\beta}\, D_b\, A \Big). \tag{8.47b}$$

We give the above transformations just to give a flavour of the type of contributions one obtains here.

One defines the vector multiplet through the superfield, $V^\dagger = V$ and

$$C = V\big|_{\theta=0=\bar\theta}, \tag{8.48a}$$

$$\phi_\alpha = i D_\alpha V\big|_{\theta=0=\bar\theta}, \tag{8.48b}$$

$$v_{\alpha\dot\alpha} = -\frac{1}{2}[D_\alpha, \bar D_{\dot\alpha}]\, V\big|_{\theta=0=\bar\theta}, \tag{8.48c}$$

$$\lambda_\alpha = iW_\alpha|_{\theta=0=\bar\theta} \qquad (8.48d)$$

and

$$D = -\frac{1}{2}D^\alpha W_\alpha|_{\theta=0=\bar\theta}. \qquad (8.48e)$$

In the above, we have used the same notatation for the auxiliary field D as well as the the covariant derivatives D and the context will obviously define which one is being taken.

Further,

$$W_\alpha = -\frac{1}{4}(\bar D_{\dot\alpha}\bar D^{\dot\alpha} - 8R)D_\alpha V \qquad (8.49a)$$

and

$$\bar W_{\dot\alpha} = -\frac{1}{4}(D^\beta D_\beta - 8R)\bar D_{\dot\alpha} V. \qquad (8.49b)$$

We note that, the projection operators of global supersymmetry given as $-(1/4)DD$ are replaced by $-(1/4)(DD-8R)$ as above, such that W_α and $\bar W_{\dot\alpha}$. in (8.49) become *spinorial* chiral and antichiral superfields. When we have *Yang-Mills* symmetry, (8.49) gets replaced by

$$W_\alpha = -\frac{1}{4g}(\bar D_{\dot\alpha}\bar D^{\dot\alpha} - 8R)e^{-gV}D_\alpha e^{gV}, \qquad (8.50)$$

with g as the coupling constant as parallel of equation (5.7) earlier. We can obtain the transformation properties of the component fields of equations (8.48); we shall not however give them here [1]. In equation (8.50), W_α gives the spinorial field strength of the gauge field as earlier, however now also including gravity.

8.6 Chiral Densities and the Lagrangian

For this purpose, we shall first consider the chiral superfield Φ. We start by using superspace variables θ which carry local Lorentz index and transform accordingly, and thus write

$$\Phi = A(x) + 2\theta^\alpha \psi_\alpha(x) + \theta^\alpha \theta_\alpha F(x). \qquad (8.51)$$

We note that the above is similar to the construction of spinors in curved space-time [3].

8.6. Chiral Densities and the Lagrangian

8.6.1 Chiral densities

We next construct $\eta^M(x,\theta)$ such that under a supersymmetry transformation,

$$\delta\Phi = -\eta^M(x,\theta)\partial_M\Phi. \qquad (8.52)$$

Explicitly, we wish to have $\eta^M(x,\theta)$ such that equation (8.52) becomes equivalent to equations (8.47). It can be seen that this is so when [1]

$$\eta^m = 2i\theta\sigma^m\bar{\xi} + (\theta\theta)\,\bar{\psi}_n\sigma^m\bar{\sigma}^n\bar{\xi}, \qquad (8.53a)$$

and

$$\begin{aligned}\eta^\alpha &= \xi^\alpha - i\,\theta\sigma^m\xi\,\psi_m{}^\alpha + (\theta\theta)\left[\frac{1}{3}M\xi^\alpha + \frac{1}{6}\,b_a\,(\epsilon\sigma^a\bar{\xi})^\alpha\right.\\ &\quad \left. - i\omega_m{}^{\alpha\beta}\,(\sigma\xi)_\beta - (1/2)\psi_n{}^\alpha\,(\psi_m\bar{\sigma}^n\sigma^m\bar{\xi})\right]. \end{aligned} \qquad (8.53b)$$

From equation (8.51), it is clear that the product of two chiral fields is a chiral field. We wish to use integration over the variables θ's to construct an invariant Lagrangian. For this purpose, we need to *define* chiral densities S as quantities transforming as

$$\delta S = -\eta^M\partial_M S - (-1)^m(\partial_M\eta^M)S. \qquad (8.54a)$$

We then easily have the result that the product of chiral density and a chiral field is a chiral density, since from (8.52) and (8.54) we obtain that

$$\delta(S\Phi) = -\eta^M\partial_M(S\Phi) - (-1)^m(\partial_M\eta^M)S\Phi. \qquad (8.54b)$$

8.6.2 Lagrangian construction

The above considerations enable us to easily construct the invariant Lagrangian. For example,

$$\text{Action} = \int d^4x\,d^2\theta\,Sg(\Phi)$$

remains invaraint under a supersymmetry transformation. We note from (8.52) and (8.54) that a chiral density is not a chiral superfield. For the construction of the Lagrangian we use the chiral density E with $e = det(e_m{}^a)$ as,

$$\begin{aligned} E &= \frac{1}{2}e + \frac{i}{2}\,e\,\theta\sigma^m\bar{\psi}_m + (\theta\theta)\left[-\frac{1}{2}e\,M^*\right.\\ &\quad \left. - \frac{1}{8}\,e\,\bar{\psi}_m\,(\bar{\sigma}^m\sigma^n - \bar{\sigma}^n\sigma^m)\,\psi_n\right]. \end{aligned} \qquad (8.55)$$

associated with the vielbein in superspace and M^*.

We now note that we may rewrite the Lagrangian of equation (5.15) with global supersymmetry and a superpotential $g(\Phi)$ as

$$\mathcal{L} = \int d^2\theta \left[-\frac{1}{8} \bar{D}\bar{D} \left(\Phi^\dagger e^{gV} \Phi \right) - \frac{1}{4} Tr(W^\alpha W_\alpha + g(\Phi) \right]$$
$$+ \ h.c. \tag{8.56}$$

In the above, we have used the same notation g for the gauge coupling constant as well as for the superpotential. In equation (8.56) above we now replace $d^2\theta$ by $d^2\theta E$, and, as already suggested, $-1/4(DD)$ by $-1/4(DD - 8R)$. This shall now yield the invariant Lagrangian for local sypersymmetry, along with some additional terms which can be present as noted below:

$$\mathcal{L} = \int d^2\theta E \left[-\frac{1}{8} \left(\bar{D}\bar{D} - 8R \right) \left(\Phi^\dagger e^{gV} \Phi \right) - \frac{1}{4} Tr(W^\alpha W_\alpha + g(\Phi)) \right.$$
$$\left. - \ \left(\bar{D}\bar{D} - 8R \right) f(\Phi^\dagger, \Phi) \right] + h.c. \tag{8.57}$$

Clearly, the sector $f(\Phi^\dagger, \Phi)$ must be neutral under the local gauge transformations so that gauge invariance is maintained. Chiral density here plays a similar role to the determinant of vierbein, or the square root of the determinant of the metric, in four-dimensional curved space time [3].

Giving a physical interpretation to the above Lagrangian for any specific problem is a nontrivial task, since the fields may be seen not to be properly normalised. We shall come back to this problem in the next chapter with an alternative approach to the same problem.

REFERENCES

1. J. Wess and J. Bagger, *Introduction to Supersymmetry*, Princeton University Press, Princeton, NJ,1983.

2. T. Eguchi, P.B. Gilkey and A.J. Hanson Phys. Rep. 66C, 213 (1980).

3. S. Weinberg, *Gravitation and Cosmology*, Wiley, New York, 1972.

4. P. West, *Introduction to Supersymmetry and Supergravity*, World Scientific, Singapore, 1986.

Chapter 9

Supergravity Lagrangian

9.1 Introduction

The construction of general form of the Lagrangian with supergravity embedded into it, as seen in the last section, is extremely complicated and has been developed over many years. We shall give here a short sketch of the concepts which have gone into this construction and shall omit most of the details involved, where the reader may look into the appropriate literature containing many facets, with ideas some of which are not really rigorous [1,2].

Before we proceed further, however, it is worthwhile to mention one particular thing: supergravity as a theory is perturbatively non-renormalisable since the Lagrangian is not a polynomial. This is quite relevant and unexpected because as we already know, supersymmetry improved renormalisablity, and thus supergravity as a generalisation of the same should do better regarding this, whereas it does just the opposite. Furthermor, supergravity Lagrangian is highly complicated, and lack of renormalisability is an effect of this. We shall thus be prepared to encounter highly cumbersome expressions of the field operators while building the supergravity Lagrangian, which is expected considering the type of expressions we had in the last chapter.

We may also remark that supergravity is a generalisation of supersymmetry. This aspect will be made obvious when we construct the supergravity Lagrangian. At the end also we shall further examine this by taking the limit of the vanishing of the gravitational interaction with a specific ansatz, so that the validity of this limit or otherwise becomes relatively clear.

9.2 Lagrangian Construction

Here we shall adopt the formalism of Cremmer et al to construct the action integral. The flat action invariant under global supersymmetry is given as (in superspace formalism, and when we permit the Lagrangian to be nonpolynomial),

$$\text{Action} = \int d^4x d^4\theta \phi(\bar{S}e^{2\bar{g}V}S) + \int d^4x \Re \int d^2\theta g(S)$$
$$+ \int d^4x \Re \int d^2\theta (W^{a\alpha}f_{ab}(S)W^b{}_\alpha), \qquad (9.1)$$

where $\phi(\bar{S}e^{2\bar{g}V}S)$ is an arbitrary function of chiral multiplet S and its complex conjugate $\bar{S}$, V is the vector superfield, and $g(S)$, called the superpotential, is an arbitrary function of S. The term denotes the flat Yang-Mills action. All the terms are taken so as to make the Lagrangian gauge invariant. When we transform the symmetry from global to local, the expressions get modified to

$$\text{Action} = \int d^4x d^4\theta E\phi(\bar{S}e^{2\bar{g}V}S) + \int d^4x d^2\theta \Big[\Re((1/R)g(S))$$
$$+ \Re((1/R)f_{ab}(S)W^{a\alpha}W^b{}_\alpha)\Big]. \qquad (9.2)$$

Here E is the superspace vielbein determinant needed for getting the invariant volume in superspace, and R is the chiral scalar curvature superfield derived from the curvature 2-form in superspace as in the last chapter in equation (8.39), where it was used to define the complex field M. In equation (9.2) a, b are gauge indices for the adjoint representation and $f_{ab}(S)$ is a function of chiral multiplets so that $f_{ab}(S)W^{a\alpha}W^b{}_\alpha$ remains gauge invaraint. One might choose $f_{ab} = \delta_{ab}$; however $f_{ab} \neq \delta_{ab}$ will be useful to give mass to the gauginos.

Thus in equation (9.2) we have three independent functions given as $\phi(\bar{S}e^{2\bar{g}V}S)$, $g(S)$ and $f_{ab}(S)$. The Lagrangian so constructed will be invariant under supergravity transformations. The expressions however are far too general to be useful, and we shall now proceed to extract possible forms for them to be tractable for the construction of supergrvity based grand unification theories. Before proceeding further, however we may note that the usual Lagrangian for gravity is not there in equation (9.2). We shall subsequently see that the Lagrangian of theory of relativity will emerge from (9.2) automatically because of local supersymmetry, which is a new feature with supergravity.

In order to get a feeling for the type of terms which emerge, we may note that from the first term in equation (9.1) we obtain [1, 2]

9.2. Lagrangian Construction

$$e^{-1}\mathcal{L}(\phi) = -\frac{1}{6}e^{-1}\phi\, \mathcal{L}_{\text{supergravity}}$$
$$+ \phi^i{}_j\left[-\frac{1}{2}D_\mu z_i\, D^\mu z^{*j} - \bar{\chi}_{Li}\, D\chi_R{}^j + \frac{1}{2}h_i h^{*j}\right]$$
$$- \phi_k{}^{ij}\left[\bar{\chi}_{Li}\,\chi_{Lj}\, h^{*k} + \chi_{Li}\, Dz_j\,\chi_R{}^k\right] + \frac{1}{2}\phi_{kl}{}^{ij}\,\bar{\chi}_{Li}\,\chi_{Lj}\,\bar{\chi}_R^k\,\chi_R^l$$
$$+ \frac{1}{3}u^*\left[\phi^i h_i - \phi^{ij}\bar{\chi}_{Li}\,\chi_{Lj}\right] + (i/3)A^\mu\left[\frac{1}{2}\phi^i{}_j\bar{\chi}_R^j\,\gamma_\mu\,\chi_{Li}\right.$$
$$\left. + \phi^i(D_\mu z_i - \bar{\psi}_\mu\,\chi_{Li})\right] + \cdots + \text{h.c.} \qquad (9.3)$$

In the above, e is determinant of vierbein, ϕ is the same function as in (9.2), the chiral multiplets S are written as (z, χ, h) with z and χ as the scalar and spinor fields and, with h as the auxiliary field. Supersripts and subscripts denote differentiation with respect to the scalar field or its complex conjugate field respectively. For example, we have

$$\phi_i{}^{jk} = \frac{\partial^3}{\partial z_i^* \partial z_j \partial z_k}\phi(z).$$

In Ref [1] besides the superscripts and subscripts, there are primes corresponding to the order of the differentiation, which we omit here. $\tilde{g}$ is the gauge coupling constant. We have used i, j for the index structure of the chiral fields and a, b as the same for adjoint gauge fields. We particularly note that the first term in equation (9.3) is the gravity Lagrangian which is not properly normalised. and, D_μ are appropriate covariant derivatives. The supergravity multiplet is specified by the vierbein, the gravitino ψ_μ, the scalar field u, same as $M(x)$ in equation (8.39), and a vector field A^μ, equivalent to "b_a" of equation (8.38). Also the gauge multiplet here is taken as $B^a{}_\mu$ for the vector field, λ^a for the gaugino field, and D^a for the auxiliary field. *The change in notation as compared to the earlier chapter of the fields and indices may be noted*, which was thought desirable since for details one has to refer to the references [1,2].

The contribution from the second term in equation (9.2), the superpotential term, is a little simpler and becomes

$$e^{-1}\mathcal{L}(g) = -\frac{1}{2}g^{ij}\bar{\chi}_{Li}\,\chi_{Lj} + \frac{1}{2}g^i h_i + \frac{1}{2}gu + \frac{1}{2}\bar{\psi}_{R\mu}\gamma^\mu \chi_{Li}\, g^i$$
$$+ \frac{1}{2}g\bar{\psi}_{R\mu}\,\sigma^{\mu\nu}\psi_{R\nu} + \text{h.c.} \qquad (9.4)$$

whereas the contribution of the gauge part of equation (9.3) becomes

$$e^{-1}\mathcal{L}(f_{ab}) = \frac{1}{2}f_{ab}\left[-\frac{1}{4}F^a{}_{\mu\nu}F^{b\mu\nu} - \frac{1}{2}\bar{\lambda}^a\gamma^\mu D_\mu\lambda^b + \frac{1}{2}D^a D^b\right.$$

$$+ \frac{1}{4} F^a{}_{\mu\nu} \bar{\psi}_\rho \sigma^{\mu\nu} \gamma^\rho \lambda^b + \frac{1}{4} F^a{}_{\mu\nu} \tilde{F}^{b\mu\nu} - \frac{1}{2} D_\mu(\bar{\lambda}^a_L \gamma^\mu \lambda^b{}_R) \Big]$$

$$+ \frac{1}{2} f_{ab}{}^i \Big[\bar{\chi}_{Li}(-\sigma.F^a + iD^a)\lambda^b{}_L - \frac{1}{2} h_i \bar{\lambda}^a_L \lambda^b{}_L$$

$$- \frac{1}{2} \bar{\psi}_{R\mu} \gamma^\mu \chi_{Li} \, \bar{\lambda}^a_L \lambda^b_L \Big]$$

$$+ \frac{1}{4} f_{ab}{}^{ij} \bar{\chi}_{Li} \chi_{Lj} \, \bar{\lambda}^a_L \lambda^b_L + h.c. \qquad (9.5)$$

In equation (9.5) the vector multiplet V^a is written as $(B^a{}_\mu, \lambda^a, D^a)$ in Wess-Zumino gauge, as stated earlier.

We next note that we can absorb the two functions $\phi(\bar{S}, S)$ and $g(S)$ into a single function

$$\phi'(\bar{S}, S) = -\frac{1}{3}\phi(\bar{S}, S)/\big[g(\bar{S})g(S)/4\big]^{1/3} \qquad (9.6)$$

and consider the single function $G(\bar{z}, z)$ given by

$$\phi'(\bar{z}, z) = exp(\frac{1}{3} G(\bar{z}, z)). \qquad (9.7)$$

The Lagrangian density will be finally given in terms of the function $G(\bar{z}, z)$ and the function $f_{ab}(z)$. *We shall omit the prime in ϕ' in our later discussions.*

The derivation of the Lagrangian density next corresponds to the following qualitative steps. Firstly, the auxiliary fields h's and D's of the chiral and vector fields are eliminated. It is then noticed that the kinetic terms associated with almost all the fields are different from what one has for the free fields. Hence the fields are to be rescaled. This is known as Weyl rescaling where it had originally been done for gravity, and is given here as

$$e_{m\mu} \to e^\sigma e_{m\mu}, \qquad (9.8a)$$

$$\lambda \to e^{-3\sigma/2} \lambda, \qquad (9.8b)$$

$$\chi \to e^{-\sigma/2} \chi, \qquad (9.8c)$$

and

$$\psi_\mu \to e^{\sigma/2} \psi_\mu. \qquad (9.8d)$$

In the above, we have σ defined as

$$e^{-2\sigma} = -\frac{1}{3}\phi. \qquad (9.9)$$

9.2. Lagrangian Construction

Another notation is found to be useful, given by

$$J = 3\,log(-\frac{1}{3}\phi), \qquad (9.10)$$

where J is known as the "Kahler potential". We may note that we are taking $\phi = \phi(\bar{z}, z)$ as a function only of the scalar fields in the above, and not of superfields. We shall be using (9.10) for the definition of the Lagrangian density. However, one still has one operation left. Even after Weyl rescaling, the quadratic expressions associated with the fermions will not be diagonal, and thus there will be some mixing. We can take into account this by the substitution [2]

$$\psi_{\mu L} \to \psi_{\mu L} - \gamma^\mu\,\frac{\phi_i}{\phi}\,\chi^i{}_R \qquad (9.11)$$

We can now write down the Lagrangian. On simplification, the bosonic Lagrangian is given as

$$\begin{aligned}e^{-1}\mathcal{L}_B =\;& exp(-G)(3 + G_k(G^{-1})_l{}^k G^l) \\ & -\frac{1}{2}g\Re\,(f^{-1})_{ab}\,(G^i T^a{}_i{}^j\,z_j)(G^k T^b{}_k{}^l z_l) - \frac{1}{2}R + G_j{}^i\,D_\mu\,z_i\,D^\mu\,\bar{z}^j \\ & -\frac{1}{4}\Re f_{ab}\,F^a{}_{\mu\nu}\,F^{b\mu\nu} - \frac{1}{4}i\,\Im f_{ab}\,F^a{}_{\mu\nu}\,\tilde{F}^{b\mu\nu}\end{aligned} \qquad (9.12)$$

The notations here are the following. With G as given from equations (9.6) and (9.7), we have as before, for example, taken

$$G^k = \frac{\partial G}{\partial z_k}$$

and

$$G_l{}^k = \frac{\partial^2 G}{\partial z_k \partial \bar{z}^l}.$$

G^{-1} is the inverse of the matrix obtained from the second order differentiation with respect to the scalar fields and their complex conjugate fields z_i and $\bar{z}^i$. Here, the first term in equation (9.12) contains the potential due to matter sector, and the second term, the same due to gauge sector, being the generalisations of F and D type terms referred to earlier in chapter 4. The third term is the gravity part of the Lagrangian. Fourth and fifth terms are extra terms which arise since we have not taken $f_{ab} = \delta_{ab}$.

We shall divide the fermionic part of the Lagrangian into two parts. Firstly, we have the kinetic part given as

$$\begin{aligned}
e^{-1}\mathcal{L}_{\text{FK}} = &\ \frac{1}{2}\Re f_{ab}\Big[-\frac{1}{2}\bar{\lambda}^a D\lambda^b + \frac{1}{2}\bar{\lambda}^a \gamma^\mu \sigma^{\rho\sigma} \psi_\mu F^b{}_{\rho\sigma} \\
&\ -\frac{1}{2}\bar{\lambda}^a_L \gamma^\mu \lambda^b{}_R\, G^i D_\mu z_i\Big] - \frac{1}{8}i\Im f_{ab}\, e^{-1} D_\mu(e\bar{\lambda}^a \gamma_5 \gamma^\mu \lambda^b) \\
&\ -\frac{1}{2}f^i{}_{ab}\, \bar{\chi}_{iL}\, \sigma.F^a \lambda^b{}_L - \frac{1}{4} e^{-1} \bar{\psi}_\mu \gamma_5 \gamma_\nu D_\rho\, \psi_\sigma\, \epsilon^{\mu\nu\rho\sigma} \\
&\ +\frac{1}{8} e^{-1} \epsilon^{\mu\nu\rho\sigma} \bar{\psi}_\mu \gamma_\nu \psi_\rho G^i D_\sigma z_i - G_i{}^j \bar{\psi}_{\mu L} D_\rho \gamma^\rho \bar{z}^i \gamma^\mu \chi_{jL} \\
&\ -\bar{\chi}_{iL}\, D_\mu \gamma^\mu z_j\, \chi^k{}_R (G_k{}^{ij} + \frac{1}{2} G_k{}^i G^j) \\
&\ + G_j{}^i \bar{\chi}_{iL}\, D_\mu \gamma^\mu \bar{\chi}^j_R + \text{h.c.} & (9.13)
\end{aligned}$$

Also, the terms of the Lagrangian from which we shall be able to obtain the masses of the fermions and which do not contain differentiation of fermions are given as

$$\begin{aligned}
e^{-1}\mathcal{L}_{\text{FM}} = &\ exp(-G/2)\Big[\bar{\psi}_{\mu R}\, \sigma^{\mu\nu} \psi_{\nu R} + \frac{1}{4} G^l (G^{-1})_l{}^k f^*_{ab,k}\, \bar{\lambda}^a_R\, \lambda^b{}_R \\
&\ + (G^{ij} - G^i G^j - G^l (G^{-1})_l{}^k G_k{}^{ij})\bar{\chi}_{iL}\, \chi_{jL}\Big] \\
&\ - \frac{i}{2} g G^i T^a{}_i{}^j z_j\, \bar{\psi}_{\mu L} \gamma^\mu \lambda^a{}_R - exp(-G/2) G^i \bar{\psi}_{\mu R}\, \gamma^\mu\, \chi_{iL} \\
&\ + \frac{i}{2}\Re(f^{-1})_{ab}\, f^k{}_{bc}\, g G^i T^a{}_i{}^j z_j\, \bar{\chi}_{kL}\, \lambda^c{}_L \\
&\ + 2ig G_i{}^j\, T^a{}_j{}^k z_k\, \bar{\lambda}^a_R\, \chi^i{}_R + \cdots + \text{h.c.} & (9.14)
\end{aligned}$$

The above is the most general form of the supergravity Lagrangian, where in this equation we have omitted writing a few terms.. We may obviously note that this Lagrangian is highly nonlinear, and the question of its perturbative renormalizability does not arise. Thus, hope of finiteness as a result of possible improved renormalisation characteristic of supersymmetry is completely reversed. One hopes that supergravity in ten dimensions as arising from superstring theories shall be finite, and thus the "complete" theory will be renormalizable and shall have predictability with a calculation to "all orders" if we include cut-off due to onset of high mass particles at Planck scale or due to compactification. Such a hope, however, is at present yet to be realised.

9.3 Identification of the terms

We shall identify the terms of the Lagrangian constructed as above in the global supersymmetric limit. The terms of the Lagrangian as in equation (9.12) shall first be analysed. It shall be instructive to take this limit of $\kappa \to 0$ in the above expressions, and show that the

9.3. Identification of the terms

supergravity theory in fact in this limit goes to the supersymmetric theory. We shall demonstrate this with a flat Kahler metric.

We note that in the above statements, we have chosen our units such that $\kappa=1$ as the choice of mass scale. We should thus first restore the constant κ considering appropriate dimensions, and then take the limit. We start this process from equation (9.12) for the bosonic Lagrangian. We note that from equations (9.6), (9.7) and (9.10) we obtain that

$$G = J - log(\kappa^6 |g|^2/4). \qquad (9.15)$$

We may note that in the above J is dimensionless and g has the mass dimension of three, and this has been included in writing down the expression. A flat Kahler metric corresponds to the ansatz

$$J = -\frac{1}{2}\kappa^2 \, \bar{z}_i \, z_i \, . \qquad (9.16)$$

In the above we have used the fact that z_i has the mass dimension one. One then obtains that

$$exp(-G) = exp(\frac{1}{2}\kappa^2 \bar{z}^i z_i)\kappa^6 |g|^2/4. \qquad (9.17)$$

We now note that the first term on the right hand side of equation (9.12) when we include κ with proper dimensional consideration becomes

$$e^{-1}\mathcal{L}^I{}_B = e^{-G}\left(3\kappa^{-4} + \kappa^{-4} G_k (G^{-1})_l{}^k G^l\right)$$

which when we take the flat Kahler metric becomes

$$= -\frac{1}{2}\left[|g^k - \frac{\kappa^2}{2}\bar{z}^k \, g|^2 - 3\kappa^2/2|g|^2\right] \qquad (9.18)$$

$$= -\frac{1}{2}\left[|g_k|^2 + \text{terms of higher order}\right] \qquad (9.19)$$

which gives the F-term contribution of the potential. We next note that the second term of equation (9.12) becomes

$$e^{-1}\mathcal{L}^{II}{}_B = -\frac{1}{2}\kappa^{-4} \, g^2 \Re(f^{-1})_{ab}(G^i T^a{}_i{}^j z_j)(G^k T^b{}_k{}^l z_l)$$

$$= -\frac{g^2}{8}(z^\dagger T^a z)(z^\dagger T^a z). \qquad (9.20)$$

In the above we recall that g is also the gauge coupling constant, which is easily distinguished from the superpotential g from the context in which the terms occur.

We next note that the third term on the right hand side of equation (9.12) yields

$$e^{-1}\mathcal{L}^{III}{}_B = \frac{1}{2}R \qquad (9.21)$$

which is classical free gravity term.

The fourth term becomes

$$\begin{aligned}e^{-1}\mathcal{L}^{IV}{}_B &= \kappa^{-2} G_i{}^j D^\mu \bar{z}^i D_\mu z_j \\ &= \frac{1}{2}(D^\mu \bar{z}^i)(D_\mu z_i),\end{aligned} \qquad (9.22)$$

which is the gauge invariant kinetic term for the scalar fields.

It is next easy to see that the fifth and sixth terms in equation (9.12) yields the free gauge field Lagrangian as

$$e^{-1}\mathcal{L}^{V+VI}{}_B = -\frac{1}{4} F^a{}_{\mu\nu} F^{a\mu\nu}. \qquad (9.23)$$

We now consider the fermion kinetic terms of equation (9.13). With appropriate consideration of the dimensions, the first term then becomes

$$\begin{aligned}e^{-1}\mathcal{L}^{I}{}_{FK} &= \frac{1}{2}\Re f_{ab}\Big[-\frac{1}{2}\bar{\lambda}^a \gamma^\mu D_\mu \lambda^b + \frac{\kappa}{2}\bar{\lambda}^a \gamma^\mu \sigma^{\rho\sigma} \psi_\mu F^b{}_{\rho\sigma} \\ &\quad -\frac{1}{2}\bar{\lambda}^a_L \gamma^\mu \lambda^b{}_R\, G^i D_\mu z_i\Big] \\ &= \frac{1}{2}\Re\Big[-\frac{1}{2}\bar{\lambda}^a \gamma^\mu D_\mu \lambda^a + \frac{\kappa}{2}\bar{\lambda}^a \gamma^\mu \sigma^{\rho\sigma} F^a{}_{\rho\sigma} \psi_\mu \\ &\quad -\frac{1}{2}\bar{\lambda}^a_L \gamma^\mu \lambda^a{}_R (-\frac{\kappa^2}{2}\bar{z}^i + g^{-1}g^i) D_\mu z_i\Big] \end{aligned} \qquad (9.24a)$$

which with $\kappa \to 0$ approaches the value

$$\frac{1}{2}\Re\Big[-\frac{1}{2}\bar{\lambda}^a \gamma^\mu D_\mu \lambda^a - \frac{1}{2}\bar{\lambda}^a_L \gamma^\mu \lambda^a{}_R (g^{-1}g^i D_\mu z_i)\Big]. \qquad (9.24b)$$

The first term above is the gaugino kinetic term. The second and third terms, when we take the limit $\kappa \to 0$, approach zero. The fourth term becomes

$$e^{-1}\mathcal{L}^{IV}{}_{FK} = -\frac{1}{4}\bar{\psi}_\mu \gamma_5 \gamma_\nu D_\rho \psi_\sigma \epsilon^{\mu\nu\rho\sigma}. \qquad (9.25)$$

Clearly the above term is the gravitino kinetic term. The fifth term here becomes in the limit $\kappa \to 0$,

$$e^{-1}\mathcal{L}^{V}{}_{FK} = \frac{1}{8} e^{-1} \epsilon^{\mu\nu\rho\sigma} \bar{\psi}_\mu \gamma_\nu \psi_\rho g^{-1} g^i D_\sigma z_i. \qquad (9.26)$$

The sixth term becomes zero, and the seventh term approaches

9.3. Identification of the terms

$$e^{-1}\mathcal{L}^{VII}{}_{FK} = \frac{1}{4}g^{-1}g^j \bar{\chi}^i_L \gamma^\mu D_\mu z_j \chi_{iR}. \qquad (9.27)$$

We next note the eighth term to be

$$\begin{aligned}e^{-1}\mathcal{L}^{VIII}{}_{FK} &= \kappa^{-2} G_i{}^j \bar{\chi}^i_L \gamma^\mu D_\mu \chi_{jR} \\ &\to -\frac{1}{2}\bar{\chi}^i_L \gamma^\mu D_\mu \chi_{iR}.\end{aligned} \qquad (9.28)$$

which clearly is the gauge invariant kinetic term of the chiral superfield fermions.

We may note in particular that the second term in equation (9.24b) and the terms of equations (9.26) and (9.27) are nonpolynomial contributions and do not occur in global supersymmetry, but have arisen while taking the limit of $\kappa \to 0$ in supergravity.

We next note the terms of the Lagrangian corresponding to equation (9.14). The first term here is given as

$$\begin{aligned}e^{-1}\mathcal{L}^I{}_{FM} &= e^{-G/2} \kappa^{-1} \bar{\psi}_{\mu R} \sigma^{\mu\nu} \psi_{\nu R} \\ &= \frac{1}{2} e^{\kappa^2 z^\dagger z/4} \kappa^2 |g| \bar{\psi}_{\mu R} \sigma^{\mu\nu} \psi_{\nu R}.\end{aligned} \qquad (9.29)$$

The above corresponds to the mass term of the gravitino, and vanishes in the limit of κ approaching zero. Most of the other terms here also approach zero. However, the third term becomes

$$\begin{aligned}&e^{-1}\mathcal{L}^{III}{}_{FM} \\ &= \kappa^{-3} exp(-G/2) \Big[G^{ij} - G^i G^j - G^l (G^{-1})_l{}^k G_k{}^{ij}\Big] \chi^i_L \chi_{jL} \\ &\to -|g|/g g^{ij} \bar{\chi}_{iL} \chi_{jL}.\end{aligned} \qquad (9.30)$$

Clearly the above corresponds to the Yukawa interactions of fermions with mesons when the superpotential g is cubic.

We have identified the terms of the supergravity Lagrangians in the limit of $\kappa \to 0$ since the terms of the Lagrangian are highly complicated, and one needs to gain some insight. The main observation we would like to make however is that we have now here completely lost any hope of perturbative renormalisability. And there is no other form of renormalisation that one can hope would be applicable here. The beauty (?) of the situation is that so many terms arose from relatively few ansatz.

In the next Chapter we shall consider some applications of the above, where really the first term of equation (9.12) will be utilised as the potential to consider vacuum structure and spontaneous symmetry breaking. The possible generalisations of a theoretical nature in the context of superstring theory which could even supply a consistent unification of all forces with renormalisability will be dealt with later. We shall only remark here that the high hopes that arose with the advent of superstring are yet to be realised, which however could be a temporary phase.

REFERENCES

1. E. Cremmer, B. Julia, J. Scherk, P. Van Nieuwenhuizen, S. Ferrara and L. Girardello, Phys. Lett. 79B, 231 (1978); Nucl. Phys, B105 (1979).

2. E. Cremmer, S. Ferrara, L. Girardello and A. Van Proyen, Phys. Lett. 116B, 219 (1982); Nucl. Phys. B212, 413 (1983).

3. Snigdha Mishra, Ph.D. thesis (1988).

Chapter 10

Model Building in Supergravity

10.1 Introduction

In this chapter we shall give a flavor of some developments of grand unified theories based on N=1 supergravity. In the previous chapter, we have given explicit construction of the supergravity Lagrangian which we shall use here for model building. Supergravity models have basically two distinct types of approaches. Sometimes one considers supergravity at low energy with soft symmetry breaking terms. This ensures the stability of hierarchy, since the symmetry breaking terms are soft and directly links with phenomenology around weak scale. The second approach is through grand unification theories, where soft breaking terms are derived through a spontaneous symmetry breaking at the Planck scale. We shall use the later approach, where the super Higgs mechanism is triggered by gravitational interactions at the tree level. In late 1982, it was realized in many almost parallel attempts [1,2] that local supersymmetry can in fact cause the breaking of Salam-Weinberg symmetry, which is associated with the extraordinary conclusion that phase transition at Planck scale would have observable effects at scales as low as 100 GeV, interrelating physics over sixteen orders of magnitude in distance or energy scales! We shall illustrate this phenomenon with the grand unification groups $SU(5)$ or $SO(10)$, where the later has interesting physics at intermediate energy scales as such, a feature which some times gets accentuated with gravitational interactions.

We shall now note some relevant terms of the supergravity Lagrangian and quote the basic formalisms for model building. As mentioned in the last chapter the scalar potential is given as

$$\mathcal{V} = \frac{1}{2} e^{\frac{1}{2}\kappa^2 |\phi|^2} \left[|g_{,\phi} + \frac{1}{2} \kappa^2 \phi^\dagger g|^2 - \frac{3}{2} \kappa^2 |g|^2 \right] + \frac{1}{2} |D_\alpha|^2, \quad (10.1)$$

where ϕ is the generic scalar field, D_α is the gauge auxiliary field and g is the gauge invariant superpotential. Also we note that summation over the repeated field ϕ as well as over the repeated gauge index α is understood. Here in the limit $\kappa \to 0$, the potential $\mathcal{V}$ reduces to potential for global supersymmetry with the usual F and D terms. The superpotential 'g' consists of two different sectors, a hidden sector responsible for spontaneous breaking of supergravity through a Polonyi singlet [3] and an "observable" sector containing quarks, leptons and Higgs supermultiplets. The two sectors communicate through gravitational interaction. In the subsequent sections, we shall discuss some phenomenologically realistic models depending upon the various possibility of the gauge groups, e.g. $SU(5)$ or $SO(10)$.

10.2   SU(5)

Here we shall quote the results of Pran Nath et al [1] where the gravitational interaction triggers the weak symmetry breaking in an $SU(5)$ model. Let

$$g = g_1(\phi) + g_2(Z) \qquad (10.2)$$

where, parallel to equation (6.13) earlier for global supersymmetry,

$$g_1(\phi) = \lambda_1 \left[\frac{1}{3} Tr\Sigma^3 + \frac{1}{2} MTr\Sigma^2\right] + \lambda_2 H'_x \left[\Sigma_y{}^x + 3M'\delta_x{}^y\right] H^y$$
$$+ \lambda_3 U H'_x H^x + \epsilon_{uvwxy} H^u M^{vw} m_1 M^{xy} + H'_x M^{xy} m_2 M'_y. \qquad (10.3a)$$

and

$$g_2(Z) = m^2(Z + B_0). \qquad (10.3b)$$

g_1, g_2 are respectively the superpotentials in the observable and hidden sectors respectively. Here $\Sigma_y{}^x$ is in the adjoint representation 24 of $SU(5)$, H^x and H'_x are in the 5 and 5^* representations, m_1 and m_2 are matrices in the generation space, and the generation index is suppressed. In the hidden sector B_0 and m are constants of dimensions of mass. The breaking pattern shall be given as, with SUSY as being short notation for supersymmetry,

$$SU(5) \times SUSY \xrightarrow[\Sigma_y{}^x]{M_U} SU(3)_C \times SU(2)_L \times U(1)_Y \times SUSY$$

$$\xrightarrow[U, H, H']{M_W} SU(3)_C \times U(1)_{em} \qquad (10.4)$$

10.2. SU(5)

Supersymmetry is preserved at the first stage of breaking governed by $\Sigma_y{}^x$. The fields H, H', U break the Salam-Weinberg symmetry as well as supersymmetry at the weak scale via super Higgs mechanism with $g_2(Z)$. Here it may be noted that the singlet Z takes the vacuum expectation value of the order of Planck scale which triggers both the symmetry breaking simultaneously as above at the much lower scale (i.e. the weak scale) as will be seen subsequently.

The potential with the observable sector g_1 superpotential has a minimum when the vacuum expectation value (vev) for H^x, H'_x, U, M^{xy} and M'_x vanish and the $\Sigma_y{}^x$ field has the vacuum expectation value.

$$\Sigma_y{}^x = 0, \qquad (10.5a)$$

$$\Sigma_y{}^x = \frac{M}{3}\left[\delta_y{}^x - 5\delta_y{}^5\delta_5{}^x\right], \qquad (10.5b)$$

or,

$$\Sigma_y{}^x = M\left[2\delta_y{}^x - 5(\delta_y{}^4\delta_4{}^x + \delta_y{}^5\delta_5{}^x)\right]. \qquad (10.5c)$$

The above are the same as equations (6.14) and imply that the vacuum is degenerate.

We shall now consider the minimization of the scalar potential where we shall note the salient extra features of the gravitational effects due to supergravity, and check the consistency of the same with the symmetry breaking pattern. The minimization condition gives, with, as before, ϕ being a generic scalar field,

$$g_{,\phi} + \frac{1}{2}\kappa^2\,\phi^\dagger\,g = 0. \qquad (10.6a)$$

We note that the model is so constructed that the D-term of the potential vanishes. The above condition when substituted in the scalar potential, gives that for appropriate symmetry breaking,

$$V_{\min} = -\frac{3}{4}\kappa^2|g(\phi)|^2 exp((1/4)\kappa^2|\phi|^2). \qquad (10.6b)$$

Here it may be noted that the vacuum degeneracy of the global supersymmetry case is lifted due to $O(\kappa^2)$ corrections in (10.6). When one introduces these corrections in the V, one thus shows that the vacuum degeneracy is lifted and the last solution corresponding to $SU(5) \rightarrow SU(3)_C \times SU(2)_L \times U(1)_Y$ is a stable one [1]. Thus the gravitational interaction removes the vacuum degeneracy of global supersymmetry, but at this stage it has exact local supersymmetry and an exact $SU(3)_C \times SU(2)_L \times U(1)_Y$ gauge invariance.

As was mentioned before, the hidden sector superpotential $g_2(Z)$ acts as a driving term in the low energy Higgs sector. The extremisation of the potential for the above Polonyi part yields

$$Z_0 = (\sqrt{2}a + \sqrt{6}b)\kappa^{-1}, \quad B_0 = (2\sqrt{2} + \sqrt{6}b)\kappa^{-1}, \qquad (10.7a)$$

$$a^2 = b^2 = 1 \text{ and } ab = -1. \qquad (10.7b)$$

Substituting this triggering term in the full scalar potential and on minimization, it now gives

$$\Sigma_y{}^x = M(1+\epsilon_1)diag[2,2,2,-3+\epsilon_2,-3-\epsilon_2], \qquad (10.8a)$$

$$U = -\frac{(\kappa m^2 x a)}{\sqrt{2}\lambda_3}, \qquad (10.8b)$$

$$H'_x = H^x = \frac{y\kappa m^2}{\sqrt{2}\lambda_3 \delta_5{}^x}, \qquad (10.8c)$$

where

$$\epsilon_1 = -\frac{\kappa m^2 a}{\sqrt{2}\lambda_1 M}, \qquad (10.9a)$$

$$\epsilon_2 = -\frac{\kappa^2 m^4 y^2 \lambda_2}{(20\lambda_1 \lambda_3^2 M^2)}, \qquad (10.9b)$$

with x and y as determined through the simultaneous equations

$$x^2 + x(3 + ab\sqrt{3} - 6\lambda) + y^2 + (1 - 3\lambda)^2 = 0, \qquad (10.10a)$$

$$y^2(3 + ab\sqrt{3} - 6\lambda) + 2xy^2 + x = 0. \qquad (10.10b)$$

In the above, $\lambda = \lambda_2/\lambda_1$. We may recall that the above equations are parallel of equations (6.19a) and (6.19b) with p and q replaced by x and y respectively. Earlier, supersymmetry was broken by hand. Here the symmetry breaking is induced by the gravitational interaction and is obtained through the extremisation of the supergravity potential. Equations (10.7) and (10.8) and (10.9) along with equations (10.10) are in fact derived by considering the extremisation equation given in (10.6) and then expanding the same in powers of the parameter $\xi = \kappa m$, where $\kappa^{-1} = (8\pi G)^{-1/2} \approx 2.4 \times 10^{18}$ GeV is the gravitational constant. From the experimental value of the weak scale symmetry breaking as given in equation (10.7c), we may see that $\xi \approx 10^{-8}$ so that making an expansion in terms of this small parameter and equating terms for the individual powers of ξ for consistency will be justifiable. This small parameter in fact links up the physics of Planck scale with that of the weak scale in a quantitative manner and makes a phase transition at Planck scale as in equations (10.9) through the Polonyi potential have observable effects of symmetry breaking in the weak scale. We note that the mass parameter m in Polonyi potential is so taken that $m\xi = \xi^2 \kappa^{-1}$ is in the weak scale, whereas the vacuum expectation value of Z as in equations (10.7) is in the Planck scale.

10.3 SO(10) models

We may also consider the grand unification gauge group SO(10) which has a richer structure in the sense of various possible intermediate symmetries. This has the advantage of including the minimal $SU(5)$ group [4] as well as the semisimple left right symmetric groups with quark lepton unification [5]. $SO(10)$ group structure with respect to global supersymmetry has been e.g. considered in Ref. [6]. However we shall first examine here an alternative version of $SO(10)$, where there is a single stage symmetry breaking with two neutral Z bosons.

10.3.1 Single Stage Symmetry Breaking

We shall consider here a single stage symmetry breaking given as

$$SO(10) \to SU(2)_L \times U(1)_R \times U(1)_{B-L} \times SU(3)_C \equiv G_{2113}$$

with supersymmetry being preserved [7]. Here the symmetry group G_{2113} becomes the electroweak unification group and thus replaces the Salam-Weinberg group $SU(2)_L \times U(1)_Y \times SU(3)_C \equiv G_{213}$ as the electroweak symmetry group. The breaking of G_{2113} to the low energy group $SU(3)_C \times U(1)_Q (\equiv G_{31} = G_{LE})$ is as earlier induced by supergravity. Due to the presence of the extra $U(1)$ symmetry the model obviously postulates an extra Z-boson, and thus is different from the standard model. As opposed to the conventional picture of $SO(10)$ symmetry, we shall see that here left-right symmetry in the charged sector is lost in the grand unification scale itself.

We shall carry out the above proposal by using the 210 representation of $SO(10)$ which shall give $SO(10) \to G_{2113}$ and still preserve global supersymmetry [8]. Then we use supergravity so that through super Higgs effects supergravity in the visible sector is broken at the weak scale with usual W and Z bosons, and, a second Z boson with mass ≤ 1 TeV. As usual we now proceed to construct the super potential, which consists of a part g_1 which breaks local gauge symmetry without breaking supersymmetry, and a part g_2 which is a Polonyi type [3] superpotential. The superpotential is $g(\Phi) = g_1 + g_2$ where Φ is a generic chiral superfield, We explicitly take

$$\begin{aligned} g_1 &= \lambda_1 \left[\frac{M}{2 \times 4!} T^{ijkl}T^{ijkl} - \frac{1}{3}\left(\frac{1}{2}\right)^4 T^{ijkl}T^{klmn}T^{mnij} \right] \\ &+ \lambda_2 \bar{\psi}_1 \left[\frac{1}{2^4 \times 4!} \Gamma^i \Gamma^j \Gamma^k \Gamma^l T^{ijkl} - aM \right] \psi_1 \\ &+ \lambda_3 \bar{\psi}_2 \left[\frac{1}{2^4 \times 4!} \Gamma^i \Gamma^j \Gamma^k \Gamma^l T^{ijkl} - a'M \right] \psi_2 \\ &+ \lambda_4 U(\bar{\psi}_1 \psi_1 + \bar{\psi}_2 \psi_2), \end{aligned} \qquad (10.11)$$

and

$$g_2(Z) = m^2(Z + B_0), \qquad (10.12)$$

the same as (10.3b). In the above i, j, k, l, m, n are $SO(10)$ indices running from 1 to 10, Z is the Polonyi singlet, and, B_0 and m are as earlier constants of dimensions of mass. T^{ijkl} is a 210 dimensional totally antisymmetric tensor, ψ_1 and ψ_2 are the 16 and 16* spinor representations of $SO(10)$ and, Γ's are the corresponding 16×16 Dirac matrices of $SO(10)$ for the respective representations. We note that there are two constants a and a' in second and third terms of equation (10.11) which we shall determine later. As was done earlier we now construct the potential as in equation (10.1) and extremise it for a variation of some of the fields with a specified ansatz. We first note that our construct is such that the D-term vanishes, since we are taking both 16 and 16* representations. We substitute $\kappa M = \xi_0$ and, as before, $\kappa m = \xi$. M will give the grand unification scale and as mentioned at the end of section 10.2, ξ will be the expansion parameter to examine symmetry breaking at successively lower scales. We shall now explicitly examine the following scenario for symmetry breaking:

$$SO(10) \times SUSY$$

$$\rightarrow \text{(vevs of } T) \rightarrow SU(2)_L \times U(1)_R \times U(1)_{B-L} \times SU(3)_C \times SUSY$$

$$\rightarrow \text{(vevs of } \psi_1, \psi_2, U) \rightarrow U(1)_Q \times SU(3)_C,$$

where the first stage of symmetry breaking will occur at order one and the second stage, at order ξ. m is taken such that $m\xi$ is in the weak scale. We note that the extremisation equation for terms of order one which also maintains global supersymmetry[8] yields that

$$\xi_0 T^{ijkl} + \frac{1}{2} T^{ijmn} T^{mnkl} = 0. \qquad (10.13)$$

10.3. SO(10) models

We then obtain the straightforward solution

$$T^{1234} = T^{1256} = T^{1278} = T^{1290} = T^{3456} = T^{3478}$$

$$= T^{3490} = T^{5678} = T^{5690} = T^{7890} = -\tfrac{1}{3}\xi_0, \qquad (10.14)$$

This yields that $SO(10) \to G_{2113}$. The above set of nonzero vevs belong to the three representations (1,1,1), (1,1,15), (1,3,15) of $SU(2)_L \times SU(2)_R \times SU(4)_C (\equiv G_{224})$ with the $SU(4)_C$ being the quark lepton symmetry group of Pati and Salam [5]. The extremisation at order one for the Polonyi part $g_2(Z)$ yields the same equation as (10.7) for the case of $SU(5)$ of section 10.2. Next, the vacuum expectation values at the weak scale are now postulated to have the form [7]

$$<U> = -\xi^2 x a/(\sqrt{2}\lambda_4)\kappa^{-1}, \qquad (10.15)$$

$$<\psi_2^0> = <\bar{\psi}_2^0> = \xi^2 z/(\sqrt{2}\lambda_4)\kappa^{-1}, \qquad (10.16)$$

and

$$<\psi_1^5> = <\bar{\psi}_1^5> = \xi^2 y/(\sqrt{2}\lambda_4)\kappa^{-1}. \qquad (10.17)$$

In the above, 0 and 5 stand for the singlet and the neutral component of the quintet in $SU(5)$ of the $SO(10)$ spinor representation, and the unknown parameters x, y, and z will get determined by considering the extremisation equations. We take the extremisation over the T's which have correction terms [7], as well as extremisation over U, ψ_1^5 as well as ψ_2^0 which give equations similar to (10.10) for the determination of the parameters x, y and z [7]. The above vacuum expectation values break $G_{2113} \times SUSY$ to the low energy group $SU(3)_C \times U(1)_Q$.

We now consider the low energy neutral current phenomenology which differs from that of the standard model. It arises from [9]

$$H_{\text{int}} = g_L W_{3L} J_{3L} + g_R W_{3R} J_{3R} + g_c C_{(B-L)/2} \qquad (10.18)$$

where the three gauge bosons W_{3L}, W_{3R} and C give rise to the two massive neutral vector mesons Z_1 and Z_2 and the massless photon. In equation (10.18) we have obviously suppressed the Lorentz index. The mass matrix of the two neutral massive bosons is given as [9]

$$M^2 = 2\frac{e^2}{(x_L(1-x_L))}\begin{pmatrix} A & B \\ B & C \end{pmatrix}. \qquad (10.19)$$

In the above $x_L = e^2/g_L^2$ and $x_R = 1 - x_L = e^2/g_R^2$, and the constants A, B and C are calculable in terms of the vacuum expectation values quoted in the above equations during extremisation. On elimination of the heavy fields, we then obtain the effective neutral current interaction Hamiltonian as [7,9]

$$H^{eff}{}_{NC} = 4\frac{G_F}{\sqrt{2}}(J_{ZL}{}^2 + \eta^2 J_{ZR}{}^2). \qquad (10.20)$$

In the above, $\eta = v_1/v_2 = y/z$ as in equations (10.16) and (10.17), with v_1 being the vacuum expectation value of the usual Higgs doublet of Salam Weinberg theory and v_2 the same for the second Higgs particle. J_{ZL} corresponds to the neutral current of the standard theory and thus η is a measure of fresh corrections. If the parameters are such that η is quite small the present model will have no difference from the standard model. We shall examine the phenomenology for different values of η. Before doing that, however, we shall examine the present model with respect to the calculation of the grand unification mass scale and its internal consistency as well as its consistency in the context of the large life time of proton. For this purpose we shall use the renormalisation equations.

For the given symmetry breaking channel, at the grand unification mass scale M_U for $SO(10)$, $g_L = g_{3R} = g_s = g$, and, $g_c = g_{(B-L)/2} = \sqrt{(3/2)}g$, where g_L, g_{3R}, g_c and g_s are the coupling constants for $SU(2)_L$, $U(1)_R$, $U(1)_{(B-L)/2}$ and $SU(3)_C$ respectively. We now consider the renormalisation group equations

$$\frac{dg_i}{dt} = A_i{}^{(S)} \frac{g_i{}^3}{16\pi^2} \qquad (10.21)$$

where, for $SU(N)$,

$$A_N{}^{(S)} = -3N + \sum T_M + \sum T_H \qquad (10.22)$$

with T_M and T_H being the quadratic $SU(N)$ Casimir invariants for the matter and Higgs fields supermultiplets. For the various groups considered above, and with the present scenario of light particles, we get [10]

$$A_L{}^{(S)} = -6 + 2N_G + \sum T_{LH}, \quad A_{3R}{}^{(S)} = 2N_G + \sum T_{3RH},$$

$$A_c{}^{(S)} = (4/3)N_G + \sum T_{cH}, \quad A_s{}^{(S)} = -9 + 2N_G + \sum T_{sH}.$$

We now examine the evolution equations of the four renormalisation groups between the weak scale and the grand unification scale given as

$$\alpha_L{}^{-1}(M_W) = \alpha_{GU}{}^{-1}(M_U) + (A_L{}^{(S)}/2\pi)ln(M_U/M_W), \qquad (10.23a)$$

$$\alpha_{3R}{}^{-1}(M_W) = \alpha_{GU}{}^{-1}(M_U) + (A_{3R}{}^{(S)}/2\pi)ln(M_U/M_W), \qquad (10.23b)$$

10.3. SO(10) models

$$\alpha_c^{-1}(M_W) = \alpha_{GU}^{-1}(M_U) + (A_c^{(S)}/2\pi)ln(M_U/M_W), \quad (10.23c)$$

and

$$\alpha_s^{-1}(M_W) = \alpha_{GU}^{-1}(M_U) + (A_s^{(S)}/2\pi)ln(M_U/M_W). \quad (10.23d)$$

where $\alpha_i = g_i^2/4\pi$ and $\alpha_{GU} = g^2/4\pi$. In order to utilize the above equations, we identify the known parameters at the weak scale as $\alpha = \alpha(M_W) = 1/128$, $\alpha_s = \alpha_s(M_W) = 0.12$ and $M_W = 82$ GeV. We then note that there are two possible values of the grand unification mass scale M_U^s or M_U^L depending on which of the four equations in (10.23) we use, given respectively as

$$\alpha^{-1} - (8/3)\alpha_s^{-1}$$

$$= \frac{1}{2\pi}\Big[A_L^{(S)} + A_{3R}^{(S)} + A_c^{(S)} - (8/3)A_s^{(S)}\Big] ln(M_U^{(s)}/M_W)$$

or,

$$\alpha^{-1} - (8/3)\alpha_L^{-1}$$

$$= \frac{1}{2\pi}\Big[A_L^{(S)} + A_{3R}^{(S)} + A_c^{(S)} - (8/3)A_L^{(S)}\Big] ln(M_U^{(L)}/M_W)$$

We then substitute the values of T_{iH} [7] for the Higgs multiplets $(2,0,1,1)$, $(2,0,-1,1)$, $(1,1/2,-1,1)$ and $(1,-1/2,1,1)$ of the subgroup $SU(2)_L \times U(1)_R \times U(1)_{B-L} \times SU(3)_C$ so as to calculate $M_U^{(s)}$ and $M_U^{(L)}$ to obtain that $M_U^{(s)} = 4.56 \times 10^{15}$ GeV and $M_U^{(L)} = 4.42 \times 10^{15}$ GeV, which reflects an internal consistency of this picture. further, if $M_U = 4.5 \times 10^{15}$ GeV, the life time for proton decay becomes [11] $\tau(p \to e^+ + \pi^0) \approx 1.7 \times 10^{35}$ years, so that proton is sufficiently long lived.

As mentioned, we have here two neutral Z-bosons. We may calculate the various values of the masses of these Z bosons for various values of η. To illustrate the nature of the conclusions, if $\eta = 0.1$, then $m_{Z_1} = 93$ GeV and $m_{Z_2} = 889$ GeV, if $\eta = 0.175$, then $m_{Z_1} = 92.5$ GeV and $m_{Z_2} = 509$ GeV, if $\eta = 0.25$, then $m_{Z_1} = 92$ GeV and $m_{Z_2} = 358$ GeV, and, if $\eta = 0.35$, then $m_{Z_1} = 91$ GeV and $m_{Z_2} = 259$ GeV. We thus find that a small value of η could still be experimentally acceptable! It is usually thought that an extra Z-boson would be a signature of superstrings; the above analysis gives an alternative scenario not connected with superstrings.

We have given the above example to illustrate some aspects of the technology of supergravity theories in the context of grand unification. There are always too many parameters, and as yet there is no specific direction for the models of the above type. It would be useful to measure η accurately, or observe a second Z boson, or supersymmetric partners of known particles so that we may have a conceptual as well as phenomenological handle to probe deeper into physics at small distances.

10.3.2 Models with Intermediate Scales

As mentioned earlier, one feature of $SO(10)$ symmetry group is that it admits a rich structure of intermediate scales. We shall here investigate some of them. Let us first consider the scenario

$$SO(10) \times SUSY$$

$$\rightarrow SU(2)_L \times SU(2)_R \times SU(4)_C \times SUSY \equiv G_{224} \times SUSY \quad (10.24a)$$

$$\rightarrow SU(2)_L \times U(1)_Y \times SU(3)_C \times SUSY \equiv G_{213} \times SUSY \quad (10.24b)$$

$$\rightarrow U(1)_Q \times SU(3)_C \equiv G_{13}. \quad (10.24c)$$

We can construct this scenario [12] by taking a superpotential g_1 which is a function of scalar Higgs particles belonging to the 54-plet representation of $SO(10)$ consisting of symmetric tensors of rank two s^{ij}, the 126 and 126* representations of $SO(10)$ consisting of tensors of rank five Γ^{ijklm} and $\bar{\Gamma}^{ijklm}$, and two decuplets H^i and H'^i along with a singlet U and the Polonyi superpotential $g_2(Z)$ as earlier. The first stage of symmetry breaking to (10.24a) takes place through the vacuum expectation values of s^{ij} parallel to that of Ref.[8] for global supersymmetry. We call this scenario as the Pati Salam scenario, since here the presence of the symmetry group G_{224} implies quark lepton unification through $SU(4)|_{color}$. The next stage of symmetry breaking to $G_{213} \times SUSY$ takes place through the vevs of scalars in 126 and 126* [12]. We take both the representations as above with equal vevs so that the D-term as in equation (10.1) vanishes, and as earlier extremisation equations become more tractable. The calculations are quite similar to that of section 10.2 for $SU(5)$ gauge group and we do not repeat them here [12]. As earlier, we make here an expansion in terms of the small parameter $\xi = \kappa m$, m being the mass parameter in the Polonyi superpotential. Here, with M as the grand unification mass scale, symmetry breaking of $SO(10)$ to G_{224} takes place at mass scale M, symmetry breaking of

10.3. SO(10) models

G_{224} to G_{213} occurs at mass scale $\kappa M^2 = \xi_0^2 \kappa^{-1}$ which may be three orders of magnitude less, and, the symmetry breaking of G_{213} to the low energy symmetry G_{13} along with supersymmetry breaking takes place at the weak scale of $\kappa m^2 = \xi^2 \kappa^{-1}$. Stability of the hierarchies against renormalisation effects as a result of nonrenormalisation theorems of supersymmetry is guaranteed since supersymmetry is good up to the weak scale.

10.3.3 Generation of Unification Hierarchies

In the above analysis, the intermediate scale occurred at the scale of $\xi_0^2 \kappa^{-1}$ as related to the grand unification scale, and not as related to the expansion parameter ξ which is quite small. Thus if the grand unification scale is the same as the Planck scale, as may be the case on the basis of some superstring inspired models, then ξ_0 will be unity, and thus the intermediate hierarchy will disappear. We have noted that in terms of the expansion parameter ξ, the weak scale is $\xi^2 \kappa^{-1}$. We now ask ourselves the question: Can we expand in terms of ξ and generate the intermediate scale $\xi \kappa^{-1}$ with the help of the extremisation equations of supergravity potential and still have supersymmetry up to the weak scale so that the stability of the hierarchy is maintained? This is amusing since then both the intermediate scale and the weak scale symmetry breaking are caused by the same mechanism, and the weak scale symmetry breaking becomes the next order effect as compared to the intermediate scale symmetry breaking. This has been examined in Ref. [13], and we describe the same briefly.

We consider here the superpotential

$$g = \lambda_1 \left[\frac{M}{2} \times \frac{1}{4!} T^{ijkl} T^{ijkl} + \frac{1}{3} \times \frac{1}{2^4} T^{ijkl} T^{klmn} T^{mnij} \right]$$
$$+ \lambda_2 \bar{\psi} \left[\frac{1}{4!} \times \frac{1}{2^4} \Gamma_i \Gamma_j \Gamma_k \Gamma_l T^{ijkl} - aM \right] \psi$$
$$+ \lambda_3 m^2 Z + \lambda_4 \bar{\psi} F_M U + A. \qquad (10.25a)$$

In the above as before i, j, k, l are $SO(10)$ tensor indices running from 1 to 10, T^{ijkl} correspond to 210-plet representation of $SO(10)$, ψ and $\bar{\psi}$ correspond to the 16 and 16* spinor representations of $SO(10)$, U is a singlet, and further the Polonyi superpotential is already present on the same footing as the other terms. As earlier we have taken both 16 and 16* so that with equality of vevs, the D-type terms in the potential (10.1) vanish. As before we shall construct the model so that $\xi = \kappa m \approx 10^{-8} \ll 1$, which will be our expansion parameter. We shall construct the model corresponding to the scenario

$$SO(10) \times SUSY$$

$$\to SU(2)_L \times SU(2)_R \times U(1)_{B-L} \times SU(3)_C \times SUSY$$

$$\equiv G_{2213} \times SUSY \qquad (10.26a)$$

$$\to SU(2)_L \times U(1)_Y \times SU(3)_C \times SUSY \equiv G_{213} \times SUSY$$
$$(10.26b)$$

$$\to U(1)_Q \times SU(3)_C \equiv G_{13}. \qquad (10.26c)$$

With $\kappa M = x_0$ we now notice that global supersymmetry is maintained if [8]

$$x_0 T^{ijkl} - \frac{1}{2} T^{ijmn} T^{mnkl} = 0. \qquad (10.27)$$

In addition to what was utilized in subsection 10.3.1, we note that the above equation is also satisfied when

$$<T^{1234}>=<T^{1256}>=<T^{3456}>=x_0. \qquad (10.28)$$

The above vacuum expectation values correspond to the representation (1,1,15) of G_{224} and thus leads to the symmetry breaking pattern $SO(10) \to G_{2213}$. With respect to the parameter ξ, we shall now assume the following expansion for the vacuum expectation values in different orders in units of κ^{-1}:

$$<T^{1234}>=<T^{1256}>=<T^{3456}>=x_0+x_2\xi^2+x_4\xi^4+\cdots,$$
$$(10.29a)$$

$$<\psi_0>=<\bar{\psi}_0>=y_1\xi+y_3\xi^3+\cdots, \qquad (10.29b)$$

$$<\psi_5>=<\bar{\psi}_5>=\eta_2\xi^2+\eta_4\xi^4+\cdots, \qquad (10.29c)$$

$$<Z>=Z_0+Z_2\xi^2+\cdots. \qquad (10.29d)$$

The distinction in the form of expansions assumed in equations (10.29b) and (10.29c) enables us to generate a solution with multiple hierarchies. As opposed to all the earlier scenarios, we shall use the counter term A here to ensure that the cosmological constant vanishes. We shall thus also take the expansions

10.3. SO(10) models

$$A = a_0 + a_2\xi^2 + \cdots. \tag{10.30}$$

and, for the superpotential, also take the expansion

$$g = b_2\xi^2 + b_4\xi^4 \cdots. \tag{10.31}$$

Such an expansion of g is obvious, since the gravitino shall have a mass in the same scale as in equation (10.29c), i.e. the weak scale. We shall here take the extremisation equations as, with comma denoting partial differentiation with respect to the corresponding field,

$$\begin{aligned} \mathcal{V}_{,\phi} &= \left[(g_{,\phi'})^\dagger g_{,\phi'\phi} + \frac{1}{2}[g^\dagger(\phi' g_{,\phi'\phi} - 2g_{,\phi}) + (g_{,\phi'}^\dagger \phi'^\dagger)g_{,\phi}]\right. \\ &+ (1/4)[(\phi'^\dagger\phi')g_{,\phi}g^\dagger + \phi^\dagger g^\dagger g] \\ &+ \left.\frac{1}{2}\mathcal{V}\,\phi^\dagger\right]E = 0, \end{aligned} \tag{10.32}$$

along with the vanishing of the cosmological constant given by

$$|g_{,\phi'}|^2 + \frac{1}{2}[(g_{,\phi'}\phi')g^\dagger + h.c.] + (1/4)|g|^2|\phi'|^2 - \frac{1}{2}|g|^2 = 0. \tag{10.33}$$

In the above equations we have taken $\kappa = 1$, and of course, summation over the repeated ϕ' fields is understood. With the expansions as taken above, and with the superpotential as in equation (10.25), we then obtain in the leading orders that

$$b_2 = \lambda_3 Z_0 + a_2, \tag{10.34a}$$

$$b_4 = \frac{3}{2} x_2(-\lambda_1 x_2 x_0 + 2\lambda_2 y_1^2) + \lambda_3 Z_2. \tag{10.34b}$$

Here it is to be noted that the zeroeth order expansion a_0 of A is chosen as

$$a_0 + \frac{1}{2}\lambda_1 x_0^3 = 0, \tag{10.35}$$

such that g vanishes at the global level. With the above expansions we then calculate the higher order corrections to the fields T, ψ_0, ψ_5, and Z as in equations (10.29). The lowest order analysis in the ψ channel shows that in equation (10.25) we must take $a = 3$. In order ξ^2 for both the T and ψ channels, the minimization equation (10.32) yields the equation

$$-\lambda_1 x_0 x_2 + \lambda_2 y_1^2 + \frac{1}{2}b_2 x_0 = 0. \tag{10.36}$$

The lowest order correction, of order ξ^4, for the Z channel as well as for the vanishing of the cosmological constant as in equation (10.33) yield two simultaneous equations in b_2 and Z_0 given as

$$-b_2 + \frac{1}{2}\left[c_2 + \frac{1}{2}d_0 b_2\right] + \frac{1}{4}(Z_0/\lambda_3)b_2^2 = 0 \qquad (10.37)$$

and

$$\frac{3}{4}b_2^2 x_0^2 + \lambda_3^2 + b_2\left[c_2 + \frac{1}{4}d_0 b_2 - \frac{3}{2}b_2\right] = 0, \qquad (10.38)$$

where

$$c_2 = -\frac{3}{2}b_2 x_0^2 + \lambda_3 Z_0, \qquad (10.39a)$$

$$d_0 = (3x_0^2 + Z_0^2). \qquad (10.39b)$$

The above equations lead to

$$Z_0^2 = 8 \pm 4\sqrt{3} \qquad (10.40)$$

and

$$\frac{b_2}{\lambda_3} = \frac{Z_0}{(1 - Z_0^2/4)}. \qquad (10.41)$$

We continue to use the extremisation equations (10.32) as well as the constraint equation (10.33) for the vanishing of the cosmological constant. In particular, we consider terms of the order ξ^6 in equation (10.33) for the vanishing of the cosmological constant, as well as terms of order ξ^4 for both T and ψ channels and terms of the order ξ^6 for the Z channel in the extremisation equations (10.32). We then obtain four equations for the five unknown parameters Z_2, x_2, y_2, η_2 and x_4 [13]. We thus find that these equations do not fully determine the parameters for the expansion of the vacuum expectation values, in which space there is a degeneracy. We may trace the cause of this degeneracy to the fact that both T and ψ channel extremisations in order ξ^2 gave rise to the same equation (10.36). One may thus conclude that the one-loop correction to the potential should be calculated. Such an exercise, to say the least, will be difficult. Instead, we shall simply examine the solutions putting $Z_2 = 0$, and examine the qualitative nature of the solutions for the symmetry breaking patterns, which is the best we are aiming at. We then obviously see that the extremisation equations as well as the vanishing of the cosmological constant are consistent with the right symmetry breaking scale to be $y_1 \xi \kappa^{-1}$, and the weak symmetry breaking scale to be $\eta_2 \xi^2 \kappa^{-1}$. From equation (10.31) it is also clear that

10.3. SO(10) models

gravitino mass also has the scale of $\xi^2 \kappa^{-1}$, so that in the visible sector the supersymmetry breaking scale coincides with the weak scale. The two scales have arisen from the supergravity theory as advertised in the beginning as was the purpose of our present analysis.

In order to have a better understanding of the physical situation, let us consider the masses of the different supermultiplets used in the model. The masses of the scalar particles will be given by the scalar potential as in equation (10.1). For example, substituting $Z = <Z> + Z'$, one obtains the mass of Z' to be in the weak scale; however, since this particle has no gauge interaction, it will not be easily produced. We next consider the fermion mass matrix, which from equation (9.14) is here given as

$$M_{ij} = exp(\kappa^2 \phi^\dagger \phi) \left[g_{,ij} + \frac{\kappa^2}{2} (\phi_i^\dagger g_{,j} + \phi_j^\dagger g_{,i}) + |g| \frac{\kappa^4}{2} (\phi_i^\dagger \phi_j^\dagger) \right]$$
(10.42)

where the indices i, j correspond to the different superfields, and as earlier summation over repeated ϕ is understood. We may note that the fermionic component of Z is the main part of the would be Goldstone fermion which is eaten up by the gravitino to make it massive. These features are the same as in $SU(5)$ supergravity models. We next turn to equation (10.42) for the other Higgs-fermion masses. The mass matrix in the T-sector shows that all the T-Higgs fermions are of the grand unification mass scale. We then examine the spinorial $\psi\{16\}$ sector and see that the $SU(2)$ weak doublet $\begin{pmatrix} \psi_4 \\ \psi_5 \end{pmatrix}$ has a mass of the order $\xi^2 \kappa^{-1}$, i.e. of the weak scale. The color triplet part becomes superheavy and is of the grand unification mass scale. We thus note that the doublet-triplet mass splitting occurs here in a natural way and the second hierarchy problem is absent.

Now we shall consider the masses both for scalars and fermions in the F (matter) sector. We note that all the scalar partners of F as well as U acquire masses of the same order as the gravitino mass. In the fermionic sector, the mass matrix which gives the neutrino mass is given as

$$M(\text{'neutrino'}) = \begin{pmatrix} 0 & 0 & \kappa m^2 \\ 0 & 0 & m \\ \kappa m^2 & m & 0 \end{pmatrix}.$$
(10.43a)

For this mass matrix, ν_L, ν_L^C and U stand for the rows and $\bar{\nu}_L$, $\bar{\nu}_L^C$ and $\bar{U}$ stand for the columns. Diagonalising this matrix, it is obvious that the right handed neutrino and the spin half part of U both acquire masses of the order of $m \approx 10^{11}$ GeV, and thus will not be observable within the foreseeable future. Further the zero mass ν(physical) corresponding to (10.43) is given as

$$\nu_L(\text{physical}) = \nu_L - \xi\, \nu_L^C. \qquad (10.43b)$$

It is thus interesting that here the physical neutrino has an enormously small ($\approx 10^{-8}$) right handed component! However, we have not been able to associate with it any measurable effect.

The above idea has been further extended by Mahapatra and Deo [14] with ξ as the hierarchy parameter to study the possibility of supergravity induced gauge hierarchies for the E_8 gauge group. Here they take the 248-plet adjoint representation of the E_8 symmetry group and study the symmetry breaking pattern

$$E_8 \times SUSY \rightarrow (\text{mass scale } M) \rightarrow E_6 \times SUSY \rightarrow$$

$$(\text{scale } \xi M) \rightarrow SO(10) \times SUSY \rightarrow (\text{scale } \xi^2 M) \rightarrow SU(5) \times SUSY$$

$$\rightarrow (\text{scale } \xi^3 M) \rightarrow G_{213} \times SUSY \rightarrow (\text{scale } \xi^4 M) \rightarrow G_{13}.$$

In the above, as taken earlier G_{213} is the standard Salam Weinberg group, and G_{13} is the low energy electromagnetic and the color symmetry group. The equations here also are complicated, and the method is completely parallel with expansions similar to those of equations (10.29) along with equations (10.30) and (10.31), with the difference that the expansions are made with the multiplication by ξ^4 at each successive order to generate the four stages of symmetry breaking with comparative ease. Further, in order that supersymmetry remains good until the weak scale, the expansion in equation (10.31) starts with a term with ξ^4 so that gravitino has the mass in the weak scale.

At present in grand unification type of theories we have higly disparate mass scales, running from Planck mass to probably the neutrino mass. It is possible that the above type of ideas help us to get these different scales in a unified manner with an economy of concepts as well as parameters.

10.4 Superheavy Higgs particles

As per the present situation regarding electro-weak unification, we may say that Salam-Weinberg theory is verified except for the observation of the Higgs particles, which are expected to have mass around the weak scale. We shall here consider a variation of Salam-Weinberg theory where, however, the Higgs doublet(s) may be superheavy, and still

10.4. Superheavy Higgs particles

the results of the "standard model" be valid. As per this picture, the Higgs particles will not be observed within foreseeable future. Here we shall construct the model in the context of $SU(5)$ based supergravity. However, a few preliminary observations shall be useful before the construction of the supergravity model to illustrate the general structure of the approach.

The electroweak scale is decided by the vacuum expectation value of the Higgs particle, and not its mass. However, in usual picture of symmetry breaking through a negative mass-square term and a $\lambda \phi^4$ term in the double-well potential, with λ of the order of unity, both mass and vev get related. We shall here first note that this need not be so. For this purpose, let us consider the Klein Gordon equation of the scalar Higgs field ϕ given as

$$(\Box + m_\phi{}^2)\phi = J_\phi. \tag{10.44a}$$

With $<J_\phi> \neq 0$, we immediately see that we shall have

$$<\phi> = <J_\phi>/m_\phi{}^2 \neq 0. \tag{10.44b}$$

Thus we can have $<J_\phi> = \mu^3$ with $\mu \approx 10^{11}$ GeV and $m_\phi \approx 3 \times 10^{15}$ GeV to obtain $<\phi> \approx 10^2$ GeV. The mechanism we contemplate shall be of the above type, so that the fermion condensate may be there in an intermediate scale, and the Higgs particles may have a mass in the grand unification scale, and still the vev of the Higgs particles will be in the weak scale.

We now take almost the same multiplets as in section 10.2, and write the superpotential as [15]

$$g = \lambda_1 [\frac{1}{3}Tr\Sigma^3 + \frac{1}{2}MTr\Sigma^2] + \lambda_2 H'_x \Sigma_y{}^x H^y$$
$$+ \lambda_3 M^{xy} M_x H'_y + \lambda_4 m^2 Z + A. \tag{10.45}$$

As earlier, $\Sigma_y{}^x$ breaks $SU(5)$ symmetry to Salam-Weinberg symmetry and preserves supersymmetry. Now, the presence of the term $\lambda_2 H'_x \Sigma_y{}^x H^y$ makes all the Higgs particles superheavy. We shall now obtain the scalar potential and assume that there is a condensate of fermions with the same quantum numbers as the Higgs doublet and examine whether the mechanism cited above becomes operative.

Clearly the grand unification symmetry is broken as before with the vevs given as

$$<\Sigma_y{}^x> = M\left[2\delta_y{}^x - 5(\delta_4{}^x \delta_y{}^4 + \delta_5{}^x \delta_y{}^5)\right]. \tag{10.46}$$

In contrast to the situation for equation (10.5c) earlier, the above equation now gives mass $3M$ to the $SU(2)_L$ doublets, and and a mass $2M$

for color triplets. Thus H and H' become superheavy. We shall now see how still the results of Salam-Weinberg theory will be valid with fermion condensates. For this purpose, in our extremisation procedure we must also include terms from the fermionic terms of equation (9.14). The relevant contribution will arise from

$$\mathcal{L}_F = -\frac{1}{2}exp(\frac{1}{2}\kappa^2\phi^\dagger\phi))\Big[g_{,ij} + \frac{1}{2}\kappa^2(\phi_j{}^\dagger g_{,i} + \phi_i{}^\dagger g_{,j})$$
$$+ \frac{1}{4}\kappa^2\phi_i{}^\dagger\phi_j{}^\dagger\Big]\bar{\chi}_i^C\chi_j - \frac{1}{16}(\bar{\chi}_i^C\chi_j)(\bar{\chi}^{iC}\chi^j) + h.c.. \qquad (10.47)$$

In the above, parallel to equation (9.14), i,j stand for all the chiral fields. From the above equation we next take the $SU(2)$ doublet components of the fermion condensates with appropriate quantum numbers and write the expression on the right hand side of equation (10.47) as

$$\mathcal{L}_F = -\frac{1}{2}exp(\frac{1}{2}\kappa^2\phi^\dagger\phi)\Big[H^4\ J^4 + H'_4\ J'_4\Big] - \frac{1}{16}\kappa^2(\bar{\chi}_i^C\chi_j)(\bar{\chi}_i^C\chi_j) + h.c.$$
$$(10.48)$$

where for the leading terms we substitute

$$J^4 = \lambda_2\ \bar{\chi}^C_{\Sigma_4{}^x}\ \chi_{H'_x} + \cdots \equiv \lambda_2\ (\bar{\chi}\chi)_1, \qquad (10.49)$$

$$J'_4 = \lambda_2\ \bar{\chi}^C_{\Sigma_x{}^4}\ \chi_{H^x} + \cdots \equiv \lambda_2\ (\bar{\chi}\chi)_2. \qquad (10.50)$$

We note that in the above we are considering *fermionic* partners corresponding to $\Sigma_4{}^x$, $\Sigma_x{}^4$, H'_x and H^x. We now recall that the fields H and H' are superheavy, and we shall eliminate them in favor of the condensates for using extremisation of the effective potential. This is done with the obvious replacement

$$H^4 \to (J^{4\dagger}/m_a{}^2)exp(-\frac{1}{4}\kappa^2\phi^\dagger\phi), \qquad (10.51a)$$

$$H'_4 \to (J'^\dagger_4/m_a{}^2)exp(-\frac{1}{4}\kappa^2\phi^\dagger\phi). \qquad (10.51b)$$

In equations (10.49) and (10.50) we have explicitly written down only the condensates with heavy fermion fields, so that chiral symmetry breaking is not involved when we take these condensates. The mechanism here thus is completely different from that of technicolor. We note that when the condensates have a vev, equations (10.51) imply vevs for the Higgs fields, which are damped by the masses of the Higgs particles as stated earlier in equation (10.44b). We may also note that these condensates do not give masses to the gauge bosons except as through the vevs of the Higgs particles at the weak scale.

10.4. Superheavy Higgs particles

We recall that $\xi_0 = \kappa M$ and $\xi = \kappa m$ and write the Lagrangian in terms of the condensates and the scalar fields in a dimensionless manner. As before we consider equations in each order in ξ. The effective bosonic Lagrangian becomes

$$\mathcal{L}^{eff}{}_B = -\frac{1}{2} exp\left\{\frac{1}{2}(|\Sigma_i{}^i|^2 + |\Sigma_a{}^a|^2 + \frac{\alpha^2}{9\xi_0{}^2}|\lambda_2((\bar{\chi}\chi)_1 + (\bar{\chi}\chi)_2)|^2\right\}$$
$$\times \left[\left\{|g_{,\Sigma_i{}^i}|^2 + |g_{,\Sigma_a{}^a}|^2 + \lambda_4 \xi^4\right\} + \frac{1}{2}\left\{(g_{,\phi}\phi)g^\dagger + h.c.\right\}\right.$$
$$+ \left.\frac{1}{4}|g|^2 \phi^\dagger \phi - \frac{3}{2}|g|^2\right]$$
$$- \frac{1}{8}\tilde{g}^2 \left(\frac{\alpha^2}{9\xi_0^2}\right)^2 \times \left[|\lambda_2(\bar{\chi}\chi)_1|^4 + |\lambda_2(\bar{\chi}\chi)_2|^4\right]$$
$$+ \frac{1}{4}\tilde{g}^2 \left(\frac{\lambda_2}{9\xi_0^2}\right)^4 \times \left[|\lambda_2(\bar{\chi}\chi)_1|^4 + |\lambda_2(\bar{\chi}\chi)_2|^4\right] \quad (10.52)$$

where $\tilde{g}$ is the gauge coupling constant and,

$$\alpha = exp(-\frac{1}{2}\kappa^2 \phi^\dagger \phi). \quad (10.53)$$

The effective fermionic Lagrangian becomes

$$\mathcal{L}^{eff}{}_F = \frac{1}{2}\frac{1}{9\xi_0^2}\left[(\lambda_2(\bar{\chi}\chi)_1)^2 + (\lambda_2(\bar{\chi}\chi)_2)^2\right]$$
$$- \frac{1}{16}\left[((\bar{\chi}\chi)_1)^2 + ((\bar{\chi}\chi)_2)^2\right]. \quad (10.54)$$

We now use equations (10.52) and (10.54) for the purpose of extremisation using an expansion in the small parameter ξ as earlier. We thus include ξ^2 order contributions to Σ and Z in addition to the zeroeth order contributions. For the condensates, we further substitute

$$\frac{\sqrt{\alpha}(\bar{\chi}\chi)_1}{m_a{}^2} = \eta_2 \xi^2 + \cdots, \quad \frac{\sqrt{\alpha}(\bar{\chi}\chi)_2}{m_a{}^2} = \eta'_2 \xi^2 + \cdots, \quad (10.55)$$

We also expand $A = a_0 + a_2 \xi^2 + \cdots$ as in equation (10.30) adjusted so that the cosmological constant vanishes as earlier, and take $g = b_2 \xi^2 + \cdots$ as in equation (10.31) so that gravitino has a mass in the weak scale. The expansions for the condensates has been taken as in equations (10.55) so that the conventional vacuum expectations values of the Higgs fields is repeated. We solve the extremisation equations with the above ansatz. We then see that a solution is possible provided we have

$$\frac{8\lambda_2{}^2}{9\xi_0^2} = 1, \quad (10.56)$$

which thus involves a fine tuning. The above is however in some sense pleasant because ξ_0 is the ratio of grand unification scale to the Planck scale, and gives an indication why the Yukawa coupling could be small. With the Z expansion as in equation (10.29d) we again obtain that $Z_0{}^2 = 8 \pm 4\sqrt{3}$, the same as equation (10.40). We also need the constraint $\lambda_2{}^4 > \tilde{g}^2 \alpha^4/2$ which is easily satisfied when Z_0 is large or equivalently α is small. With the substitution

$$v = \lambda_2 \eta_2 \xi^2 \kappa^{-1} \qquad (10.57)$$

we obtain that $m_W = \tilde{g}v/\sqrt{2}$ and $m_Z = m_W/cos\theta_W$ and all the results of Salam Weinberg theory remain unaltered except for the fact that the Higgs particles will not be observed, and that the radiative corrections coming from light Higgs particles will be absent until we reach a large enough scale, which could be that of condensates.

The above conclusions are based on supergravity theory which is nonrenormalisable. However, the mechanism envisaged in equation (10.44) could as well occur in a renormalisable theory, and Salam Weinberg symmetry breaking take place through the existence of nonzero vevs for condensates. We can show that as in standard Higgs mechanism, in such a case also the resulting theory with condensates will continue to be renormalisable [15].

10.5 Radiative Symmetry Breaking

In the earlier sections our calculations have been always at the tree level; i.e. the potentials which have been extremised did not include quantum corrections. Higher order corrections, in fact of a nonperturbative nature, were only utilized in equations (10.23) in order to obtain the grand unification scale from the weak scale through the renormalisation group equations. These involved the idea of the "running coupling constants" which were dependant of the energy scales. In a similar way we have also renormalisation group equations for masses of particles which can be considered in the context of spontaneous symmetry breaking.

Let us imagine the double well potential which gives rise to the Higgs mechanism for symmetry breaking. Here the "mass square" term has a negative sign, and this gives rise to symmetry breaking since it gives rise to a nonzero vev for the corresponding Higgs field. When mass square is scale dependant, it could become negative dynamically through the renormalisation group equations for the same. Masses in grand unification theories are related to the Yukawa couplings and hence the renormalisation equations for the same will also be relevant. We shall illustrate the principles involved in this mechanism with a simple model.

10.5. Radiative Symmetry Breaking

Let us consider a supersymmetric grand unification theory as noted in earlier sections. At the weak scale, the following particles are expected to be there. (i) There will be generally two Higgs doublets H^a and H'_a with $a = 1, 2$ with their supersymmetric partners; (ii) there will be the matter sector of fermions with the supersymmetric partners; (iii) there will also be Salam Weinberg gauge bosons along with the gluons. We can therefore write down the potential in terms of the above fields at this scale. Since supersymmetry gets broken at this scale, the potential will involve soft supersymmetry breaking terms, i.e. terms which break supersymmetry but do not introduce any ultraviolet divergences. There will be renormalisation group equations for the corresponding parameters, which shall be here the gauge couplings as earlier, the Yukawa couplings, and the masses of the particles. When we include all the parameters, these equations are highly complicated [16] and can not be solved. As mentioned however the Yukawa coupling parameters multiplied by the vev of the Higgs particles gives the masses of the fermions or, when supersymmetry is there, also for the scalar partners. The Yukawa coupling of the top quark must be quite large, since the top quark has not been observed so far and is believed to be quite massive. We shall assume that the renormalisation group equations are dominated by this coupling parameter and in comparison put all the other Yukawa couplings as well as the gauge couplings as zero. This is likely to be qualitatively reasonable, and we can with this approximation solve the equations analytically, which will illustrate the mechanism.

The potential in the Higgs sector can be generally written as

$$\mathcal{V}_H = m_H^2 \bar{H}^a H^a + m_H'^2 \bar{H}'_a H'_a + m_3 m_g (H'_a H^a + h.c.)$$
$$+ \frac{1}{8} g_1^2 (\bar{H}^a H^a - \bar{H}'_a H'_a)^2 + \frac{1}{8} g_2^2 (\bar{H}\tau_i H - \bar{H}'\tau_i H')^2 \quad (10.58)$$

In the above, m_g is the mass of the gravitino and indicates the soft symmetry breaking term. We also note that in the above we have not taken any singlet U that was being taken in most grand unification theories.

We next write down only the top quark sector for the matter part of the potential, since, as stated, this part will be more important for the renormalisation group equations. With $\tilde{t}_L$ and $\tilde{t}_R$ as the scalar partners of the top quark, we write the corresponding potential as

$$\mathcal{V}_{\text{top}} = h_t^2 |\tilde{t}_R \tilde{t}_L| + h_t^2 |\tilde{t}_R H|^2 + m_g A_t h_t (\tilde{t}_R H \tilde{t}_L + h.c.)$$
$$+ m_{t_R}^2 |\tilde{t}_R|^2 + m_{t_L}^2 |\tilde{t}_L|^2. \quad (10.59)$$

h_t is the Yukawa coupling for the top quark. A_t is a coupling parameter which enters into the low energy potential and depends on the grand unification theory one may choose [17]. With $\alpha_t = h_t^2/4\pi$, and with the variable $t = ln(\mu/M)$ where μ is the scale at which we are considering the equations and M is the grand unification scale, the renormalisation group equations for the scalar masses become,

$$\frac{dm_H^2}{dt} = \frac{3\alpha_t}{2\pi}\left[m_H^2 + m_{t_R}^2 + m_{t_L}^2 + A_t^2 m_g^2\right], \tag{10.60}$$

$$\frac{dm_{t_L}^2}{dt} = \frac{2\alpha_t}{2\pi}\left[m_H^2 + m_{t_R}^2 + m_{t_L}^2 + A_t^2 m_g^2\right], \tag{10.61}$$

and

$$\frac{dm_{t_R}^2}{dt} = \frac{\alpha_t}{2\pi}\left[m_H^2 + m_{t_R}^2 + m_{t_L}^2 + A_t^2 m_g^2\right]. \tag{10.62}$$

Also, the renormalisation equations for α_t and A_t become

$$\begin{aligned}\frac{d\alpha_t}{dt} &= \frac{3\alpha_t^2}{\pi} - \frac{\alpha_t}{\pi}\left[\frac{8}{3}\alpha_3 + \frac{3}{2}\alpha_2 + \frac{13}{18}\alpha_1\right] \\ &\approx \frac{3\alpha_t^2}{\pi},\end{aligned} \tag{10.63}$$

$$\frac{dA_t}{dt} = \frac{3\alpha_t}{\pi}A_t. \tag{10.64}$$

As noted earlier, in equation (10.63) we have neglected the gauge coupling constants α_2, α_1, α_3 of the gauge group G_{213}.

We next note that at $t = 0$ the appropriate boundary conditions are given as $\alpha_t = \alpha_0$, $A_t = A_0$, and, $m_H^2 = m_{t_R}^2 = m_{t_L}^2 = m_g^2$. Equations (10.63) and (10.64) are trivial to solve, and yield that

$$\alpha_t = \frac{\alpha_0}{(1-\eta)}, \text{ and } A_t = \frac{A_0}{(1-\eta)}, \tag{10.65}$$

where $\eta = (3\alpha_0 t/\pi)$. In order to solve for equations (10.60), (10.61) and (10.62), let us substitute $\phi = (m_H^2, m_{t_L}^2, m_{t_R}^2)$ and take three vectors $\phi_1 = (3,2,1)$, $\phi_2 = (1,-2,1)$, $\phi_3 = (1,0,-1)$. We can then write

$$\phi = C_1(\eta)\phi_1 + C_2(\eta)\phi_2 + C_3(\eta)\phi_3,$$

with the renormalisation group equation given as

$$\frac{dC_1}{d\eta} = \frac{C_1}{(1-\eta)} + \frac{A_0^2 m_g^2}{6(1-\eta)^2}, \tag{10.66}$$

$$\frac{dC_2}{d\eta} = \frac{dC_3}{d\eta} = 0. \tag{10.67}$$

One then easily obtains the solutions for the scalar masses as

$$m_H{}^2 = 3C_1 + C_2 + C_3 = 3m_g{}^2 f(\eta) - \frac{1}{2}m_g{}^2, \tag{10.68}$$

$$m_{\tilde{t}_L}{}^2 = 2C_1 - 2C_2 = 2m_g{}^2 f(\eta), \tag{10.69}$$

$$m_{\tilde{t}_R}{}^2 = C_1 + C_2 - C_3 = m_g{}^2 f(\eta) + \frac{1}{2}m_g{}^2. \tag{10.70}$$

Here

$$f(\eta) = \frac{1}{2(1-\eta)} \left[1 - \frac{A_0{}^2 \eta}{3(1-\eta)} \right]. \tag{10.71}$$

We note that since supersymmetry is good till the weak scale, both t quark and its scalar partner have the same mass. Clearly below the grand unification scale, $t < 0$ and thus $-\eta > 0$. Hence, $f(\eta)$ approaches zero as $-\eta$ approaches infinity so that $m_H{}^2$ becomes negative for sufficiently small μ, i.e. as low energy is approached. This will immediately imply spontaneous symmetry breaking, giving mass to the W and Z. The known value of these masses determines α_t and hence the top quark mass. With the above analysis the top quark mass appears to be as large as 180 GeV or larger than that [17]. Initially it was thought that such a value is not at all acceptable, but at present one does not know. It should be noted that if one includes the contributions from the terms of renormalisation group equations which were neglected as above and solves the problem numerically, this mass becomes model dependant and no unique value can be predicted. The obvious reason for this is that so far we have no hint regarding the scales for supersymmetric particles, as well as regarding which grand unification may preferred as compared to others.

10.6 Remarks

We note that the main objective of supersymmetry and supergravity is to unify the laws of nature to include strong, electromagnetic, weak as well as gravitational interactions in one framework. The initial enthusiasm for supersymmetry was that it enabled us to maintain stability of the hierarchies in grand unification schemes against changes due to renormalisation. The main enthusiasm for supergravity on the other hand was due to the fact that gravitational interactions could finally be included along with the other well-known interactions of quantum field theory. However, this happened with a heavy price to pay - the theory became nonrenormalisable and thus stopped being a theory at all.

In this context it is believed that supergravity might arise from a deeper theory - the superstring theory. In such a scenario, the natural cutoff for supergravity theory in the context of nonrenormalisability shall be the Planck scale, which is the scale at which infinite number of particles arise in the superstring theories. From the weak scale to the Planck scale we expect the lowest order contributions to be normally enough when the heavy modes are excluded, and in the light sector, renormalisation procedure is expected to work. This was the philosophy in the last subsection while we were using the renormalisation group equations in going from the weak scale to the grand unification scale. In general this procedure is likely to be correct. However, when the energy increases to Planck scale, we should really consider the superstring theory or, some parallel of quantum gravity.

In superstring theories the price one pays is that it is a consistent thoery only in dimensions higher than four, and one has to come to four dimensions through a direct or indirect compactification. Higher dimensional theories were originally considered by Kaluza and Klein, and in the next Chapter we shall examine the structure of such theories in the context of superstrings as well as otherwise.

As happens in physics, the unification of the laws of nature is inevitable, and in fact is our primary objective. While doing this, the theories progressively become more abstract and naturally more removed from ideas involving extrapolation from "conventional" experience or wisdom, until such time as the new theories stand the test of time, and in its turn become conventional. We have not at present achieved this status, and the material in the rest of the book is an attempt to find the possible directions to achieve the same.

REFERENCES

1. A.H. Chamseddine, R. Arnowitt and P. Nath, Phys. Rev. Lett. 49, 970 (1982).

2. R. Barbieri, S. Ferrara and C.A. Savoy, Phys. Lett. 119B, 342 (1982).

3. J. Polonyi, University of Budapest Report No. KFKI-1977-93 (1977).

4. H. Georgi and S.L. Glashow, Phys. Rev. Lett. 32, 438 (1974).

10.6. Remarks

5. J.C. Pati and A. Salam, Phys. Rev. D10, 275 (1974).

6. C.S. Aulakh and R.N. Mohapatra, Phys. Rev. D28, 217 (1983).

7. S. Mahapatra, S. Mishra and S.P. Misra, Phys. Rev. D30, 2019 (1984).

8. J. Maalampi and J. Pulido, Nucl. Phys. B228, 242 (1983).

9. V. Barger, E. Ma and K. Whisnant, Phys. Rev. D26, 2380 (1982).

10. R. Slansky, Phys. Rep. 79, 1 (1981).

11. K. Biswal, L. Maharana and S.P. Misra, Phys. Rev. D25, 266 (1982).

12. S. Mahapatra and S.P. Misra, Phys. Rev. D31, 656 (1985).

13. S. Mishra and S.P. Misra, Phys. Rev. D32, 3043 (1985).

14. S. Mahapatra and B.B. Deo, Phys. Rev. D38, 3554 (1988).

15. S. Mishra and S.P. Misra, Phys. Lett. 186B, 99 (1987).

16. C. Kounnes, A.B. Lahanas, D.V. Nanopoulos, M. Quiros, Nucl. Phys. B236, 438 (1984).

17. H.P. Nilles, Phys. Rep. 110, 1 (1984).

Chapter 11

Supergravity in Ten Dimensions

11.1 Introduction

A search for higher dimensional world in which the observed four dimensional space-time may be immersed, started with the papers of Kaluza and Klein [1]. Here the basic idea is that gauge theories arise from geometrical symmetries of higher dimensions. However the theory is not unique as one can start with any arbitrary dimension and also "fold" the extra dimensions into a small length in an arbitrary manner. On the other hand, the superstring theory is uniquely defined in ten dimensions, which can describe the gauge interactions as well as gravity. The amusing feature of this is also that the low energy limit of superstring theory describes N=1 supergravity in ten dimensions which thus includes gravitation.

Therefore in the present chapter we shall deal with ten dimensional N=1 supergravity. We shall motivate these discussions both in the context of an eleven dimensional Kaluza Klein theory, as well as based on the heterotic superstring theory as an illustration of the link up of Kaluza Klein theories with string theories.

11.2 Supergravity Lagrangian in ten dimensions

Initially, the study of dual spinor model and interacting closed strings revealed the existance of N=1 supergravity in ten dimensions [2]. However it was also established by Cremmer et al [3,4] that N=1 supergravity exists in eleven dimensions and on consistent compactification gives rise to ten dimensional Lagrangian. In Kaluza klein theories eleven dimensions is also relevant since $SU(3)_C$ can be defined over CP^2 which has four dimensions, $SU(2)_L$ can be defined on S^2 which has two di-

11.2. Supergravity Lagrangian in ten dimensions

mensions, and, definition of $U(1)_Y$ needs one more dimension and thus seven extra dimensions is the minimum to be able to absorb Salam Weinberg symmetry structure. We shall presently consider this along with its compactification to supergravity in ten dimensions because of its importance in the context of superstring based theories.

The field content of eleven dimensional supergravity are the vielbein $E_M{}^A$, Rarita-Schwinger Majorana field ψ_M, antisymmetric 3-form potential A_{MNP}. Here the index M runs for $0, 1, \cdots, 10$ as the Einstein indices, and $A = 0, 1, \cdots, 10$ as the Lorentz indices as mentioned in Chapter 8, equation (8.11). We then take a particular ansatz for the vielbein, using local SO(1,10) symmetry,

$$E_M{}^A = \begin{pmatrix} E_\mu{}^a & E_\mu{}^{10} = B_\mu \\ 0 & E_{10}{}^{10} = \phi \end{pmatrix}, \quad (11.1a)$$

where μ and a runs for $0, 1, 2, \cdots 9$ for Einstein and Lorentz indices respectively. Further the spinor ψ_M and the totally antisymmetric potential A_{MNP} decompose to

$$A_{MNP} = (A_{\mu\nu\rho}, A_{\mu\nu 10} = a_{\mu\nu}) \quad (11.1b)$$

and

$$\psi_M = (\psi_\mu, \psi_{10} = \lambda). \quad (11.1c)$$

In equation (11.1a) ϕ is a scalar and B is a gauge field as was the case in original Kaluza Klein theory [1]. In equation (11.1b) $a_{\mu\nu}$ is a 2-form potential and λ is a Dirac spinor. The details of the dimensional reduction technique is explained in Reference [4]. However, we shall omit the algebra here and only indicate the necessary assumptions involved [5]. Here we first set B_μ and $A_{\mu\nu\rho}$ to zero. Further, due to special properties of ten dimensional spinors, one can impose the Majorana and Weyl conditions on ψ_μ and λ. Thus each of the Majorana spinors split into two Majorana Weyl ones. One further constrains ψ_μ to have only left-handed component and λ having only right-handed components [5]. We now write down the $d = 10$ supergravity Lagrangian with the identified supergravity multiplet $(E_\mu{}^a, \psi_\mu, a_{\mu\nu}, \lambda, \phi)$ corresponding to vielbein, Rarita-Schwinger field, two form potential, Dirac field and a dilaton scalar field respectively to be given as [6],

$$\mathcal{L} = \frac{1}{2}eR - \frac{1}{2}e\bar{\psi}_\mu \Gamma^{\mu\rho\sigma} D_\rho(\omega)\psi_\sigma - \frac{3}{4}e\phi^{-3/2} f_{\rho\mu\nu} f^{\rho\mu\nu}$$
$$- \frac{1}{2}e\bar{\lambda}\Gamma^\mu D_\mu(\omega)\lambda - \frac{9}{16}e\left(\frac{\partial_\mu \phi}{\phi}\right)^2 - \frac{3}{8}\sqrt{2}e\bar{\psi}_\rho\left(\Gamma^\mu \frac{\partial_\mu \phi}{\phi}\right)\Gamma^\rho\lambda$$
$$+ \frac{1}{16}\sqrt{2}e(\phi)^{(-3/4)} f_{\theta\phi\rho}\left[\bar{\psi}_\mu \Gamma^{\mu\theta\phi\rho\nu}\psi_\nu + 6\bar{\psi}^\theta \Gamma^\phi \psi^\rho - \sqrt{2}\bar{\psi}_\mu \Gamma^{\theta\phi\rho}\Gamma^\mu \lambda\right]$$
$$+ \text{ four fermion couplings}. \quad (11.2)$$

Here the first term is the Einstein Lagrangian except that the construction involves some additional torsion, e denotes $det(E_\mu{}^a)$ and the two form potential $a_{\mu\nu}$ appears only through 3-form field strength $f_{\rho\mu\nu} = \partial_{[\rho}a_{\mu\nu]}$. Multiindex Γ matrices denote the antisymmetrised products of the Dirac matrices in ten dimensions, e.g. $\Gamma^{\mu\rho\sigma} = \Gamma^{[\mu}\Gamma^\rho\Gamma^{\sigma]}$ in ten dimensions. We may also note that the last term $\bar{\psi}_\mu \Gamma^{\theta\phi\rho}\Gamma^\mu \lambda$ in the Lagrangian is nonvanishing only when ψ_μ and λ are opposite handed as mentioned earlier.

We may now couple Yang-Mills theory to the gravity Lagrangian for $d = 10$. Here the 3-form field strength for pure gravity is to be modified to effect the coupling of 2-form potential $a_{\mu\nu}$ to gauge potential A_μ. For simplicity we first discuss the appropriate modification for abelian case given by [6]

$$F_{\rho\mu\nu}^{(\text{abelian})} = f_{\rho\mu\nu} - \frac{1}{\sqrt{2}}\kappa A_{[\rho}f_{\mu\nu]}, \qquad (11.3)$$

where $A_{[\rho}f_{\mu\nu]}$ is called the abelian Chern-Simons term. Here $f_{\mu\nu}$ is the *gauge field strength* to be distinguished from the three component gravitational field strength. κ is the gravitational coupling constant. $F_{\rho\mu\nu}^{(\text{abelian})}$ is gauge invariant if one postulates that the gauge transformation

$$\delta A_\mu = \partial_\mu \Lambda \qquad (11.4a)$$

is accompanied by a transformation for the 2-form

$$\delta a_{\mu\nu} = \frac{1}{2}\kappa \Lambda f_{\mu\nu}. \qquad (11.4b)$$

One then examines the supersymmetric invariance for $F_{\rho\mu\nu}^{(\text{abelian})}$. The supersymmetric transformation for the gauge field A_μ is given as

$$\delta_{\text{susy}} A_\mu = \frac{1}{2}\Omega_\mu, \qquad (11.5a)$$

where $\Omega_\mu = \phi^{3/8}\bar{\epsilon}\Gamma_\mu \chi$ with ϵ as the supersymmetric transformation parameter and χ the *gaugino* field. The corresponding transformation for $a_{\mu\nu}$ is,

$$\delta_{\text{susy}} a_{\mu\nu} = \frac{1}{\sqrt{2}}\kappa \Omega_{[\mu} A_{\nu]}. \qquad (11.5b)$$

We then calculate the supersymmetric transformation for $F_{\rho\mu\nu}^{(\text{abelian})}$ as

11.2. Supergravity Lagrangian in ten dimensions

$$\delta_{\text{susy}} F_{\rho\mu\nu}^{(\text{abelian})} = -\frac{1}{\sqrt{2}}\kappa\Omega_{[\rho} f_{\mu\nu]}, \qquad (11.6)$$

which is gauge invariant. Hence the prescription for the abelian 3-form field strength (11.3) is correct for gauge and supersymmetric invariance.

We may now generalise it to nonabelian Yang-Mills theory. In this case, one has to write down the nonabelian version of (11.3) as given by, [6]

$$F_{\rho\mu\nu}^{(\text{nonabelian})} = f_{\rho\mu\nu} - \frac{1}{\sqrt{2}}\kappa\chi_{\rho\mu\nu}, \qquad (11.7)$$

where $\chi_{\rho\mu\nu}$ is the Chern-Simons 3-form defined as,

$$\chi_{\rho\mu\nu} = Tr\left[A_{[\rho} F_{\mu\nu]} - \frac{2}{3}g A_{[\rho} A_\mu A_{\nu]}\right], \qquad (11.8a)$$

with the property,

$$\partial_{[\sigma}\chi_{\rho\mu\nu]} = \frac{1}{2} Tr\left[F_{[\sigma\rho} F_{\mu\nu]}\right]. \qquad (11.8b)$$

Here, we may note that, the corresponding identity for the abelian case is given as,

$$\partial_{[\sigma} a_{[\rho} f_{\mu\nu]]} = \frac{1}{2} f_{[\sigma\rho} f_{\mu\nu]}. \qquad (11.9)$$

We shall now examine the gauge invariance of $F_{\rho\mu\nu}^{(\text{nonabelian})}$. For an infinitesimal gauge transformation $\delta A_\mu = D_\mu \Lambda$ where D_μ is the covariant derivative, the accompanied transformation for the 2-form $a_{\mu\nu}$ is given as

$$\delta a_{\mu\nu} = \sqrt{2}\kappa Tr\left[\Lambda D_{[\mu} A_{\nu]}\right] \qquad (11.10)$$

so that $\delta F_{\rho\mu\nu}^{(\text{nonabelian})} = 0$. Further for the supersymmetry transformation, $\delta_{\text{susy}} A_\mu = \frac{1}{2}\Omega_\mu$. The corresponding transformation for $a_{\mu\nu}$ is taken as,

$$\delta_{\text{susy}} a_{\mu\nu} = \frac{1}{\sqrt{2}}\kappa Tr\left[\Omega_{[\mu} A_{\nu]}\right], \qquad (11.11)$$

in analogy with (11.5b). Hence $F_{\rho\mu\nu}^{(\text{nonabelian})}$ is transformed to,

$$\delta_{\text{susy}} F_{\rho\mu\nu}^{(\text{nonabelian})} = -\frac{1}{\sqrt{2}}\kappa Tr\left[\Omega_{[\rho} F_{\mu\nu]}\right], \qquad (11.12)$$

which is gauge invariant.

We shall now use the notations of differential geometry to rewrite equations (11.7) and (11.8) for the 3-form field strength with an appropriate normalisation as given by,

$$H = da + \omega^0_{3Y}, \qquad (11.13)$$

where the 3-form H corresponds to $H_{\rho\mu\nu}$, 2-form a to $a_{\mu\nu}$ and, the Yang-Mills Chern Simon term ω^0_{3Y} is given as

$$\omega^0_{3Y} = Tr\left[AF - \frac{1}{3}A^3\right]. \qquad (11.14)$$

Although gauge invariance is satisfied for the above, the theory does not become anomaly free. This possibility was carefully studied by Green and Schwarz [7]. Here one adds further a *Lorentz* Chern-Simons term to H, here given as,

$$H = da + \omega^0_{3Y} + \omega^0_{3L}. \qquad (11.15)$$

In anology with (11.14), one takes the Lorentz Chern Simon term ω^0_{3L} as

$$\omega^0_{3L} = Tr\left[\omega R - \frac{1}{3}\omega^3\right], \qquad (11.16)$$

with ω as the Lorentz 1-form connection and R the covariant 2-form as in equation (8.8) and (8.14) respectively. It has been shown in Ref.[7] that the above form is adequate for the cancellation of Yang-Mills anomalies for specific gauge groups $SO(32)$ or $E_8 \times E_8'$. Anomaly in ten dimensions is related to hexagonal diagrams, just as anomalies in four dimensions are related to triangular diagrams. Further, the hexagonal diagrams are related to a 12-form,

$$\Omega_{12} = Tr\left[F^6\right], \qquad (11.17a)$$

which is to be evaluated with the appropriate group structure. For example, one obtains for gravitational anomalies [7], with n as the dimension of the gauge group,

$$\Omega_{12} = -\left[\frac{1}{32} + \frac{(n-496)}{13824}\right](TrR^2)^3$$
$$- \left[\frac{1}{8} + \frac{(n-496)}{5760}\right](TrR^2)(TrR^4) - \left[\frac{(n-496)}{7560}\right]TrR^6. \qquad (11.17b)$$

The last term above gives rise to anomaly which can not be cancelled by any counter term [7]. This term vanishes if $n = 496$, as in the case

11.3. Heterotic Superstring theory

of $SO(32)$ or $E_8 \times E'_8$ gauge groups. We have quoted the expressions (11.17b) to merely give a flavor of the "magical" way in which anomaly cancellation occurs. It was demonstrated that the Yang-Mills and the mixed anomalies also vanish for the same choice of the gauge group [7]. This had given rise to the hope that the symmetry group could be fixed uniquely from basic principles.

We shall now construct the ten dimensional Lagrangian with the modified gravitational field strength $H_{\rho\mu\nu}$ as in equation (11.15), along with appropriate gauge bosons and the gauginos. These lead to a modified form of Largrangian, consisting of the bosonic and the fermionic parts as,

$$e^{-1}\mathcal{L}_B = -\frac{1}{2}R + \frac{3}{4}\phi^{-3/2}H_{\rho\mu\nu}H^{\rho\mu\nu} + \frac{9}{16}\left(\frac{\partial_\mu \phi}{\phi}\right)^2$$
$$- \frac{1}{4}\phi^{-3/4}F^\theta_{\mu\nu}F^{\theta\mu\nu}, \qquad (11.18a)$$

$$e^{-1}\mathcal{L}_F = -\frac{i}{2}\bar{\psi}_\mu F^{\mu\nu\rho}D_\nu(\omega)\psi_\rho + \frac{i}{2}\bar{\lambda}\Gamma^\mu D_\mu(\omega)\lambda$$
$$+ \frac{3}{8}\sqrt{2}\bar{\psi}_\mu(\Gamma^\rho \frac{\partial_\rho \phi}{\phi})\Gamma^\mu \lambda - \frac{\sqrt{2}}{16}\phi^{-3/4}H_{\rho\mu\nu}$$
$$\times \left[i\bar{\psi}_\sigma \Gamma^{\sigma\mu\nu\rho\eta}\psi_\eta + 6i\bar{\psi}^\mu \Gamma^\nu \psi_\rho + \sqrt{2}\bar{\psi}_\sigma \Gamma^{\mu\nu\rho}\Gamma^\sigma \lambda - i\bar{\chi}^\theta \Gamma^{\mu\nu\rho}\chi^\theta \right]$$
$$+ \frac{i}{2}\bar{\chi}^\theta \Gamma^\mu D_\mu(\omega)\chi^\theta - \frac{i}{4}\kappa\phi^{-3/8}(\bar{\chi}^\theta \Gamma^\mu \Gamma^{\nu\rho}F^\theta_{\nu\rho})(\psi_\mu + \frac{i\sqrt{2}}{12}\Gamma_\mu \lambda)$$
$$+ \text{ four fermion interactions.} \qquad (11.18b)$$

Here $F^\theta_{\mu\nu}$ is the gauge field-strength and χ^θ are the gaugino fields with θ as the gauge index in the adjoint representation, ψ_μ is the gravitino field (Rarita-Schwinger) with spin 3/2, λ is the spin 1/2 Dirac field, and $H_{\rho\mu\nu}$ is the field-strength in the gravity sector. In the next section, we shall see that this Lagrangian also corresponds to the low energy limit of heterotic string theory, where the gauge group is "uniquely" fixed due to string dynamics.

11.3 Heterotic Superstring theory

In this section, we shall discuss some fundamental aspects of heterotic string theory [8] and identify the physical zero string modes. There are two different approaches. One is a description with a right moving superstring in ten dimensions along with a left moving bosonic string in twentysix dimensions. The topological structure of a consistent torus compactification of these sixteen dimensions out of 26 then yield 496 vector modes which can be identified with $E_8 \times E_8'$ or $SO(32)$ symmetry. This method of symmetry generation is rather abstract. Instead we shall choose a description with a right moving superstring in ten dimensions and 32 fermionic fields which are scalars in space-time. These will subsequently yield the group structure. Here in *light cone gauge*, the string action is given as [9],

$$S^{L.C.} = -\frac{1}{4\pi\alpha'} \int d\sigma d\tau \left[\partial_\alpha X^m \partial^\alpha X^m + i\bar{S}\Gamma^-(\partial_\tau + \partial_\sigma)S \right.$$
$$\left. + \{\bar{\psi}^A(\partial_\tau - \partial_\sigma)\psi^A + \bar{\psi}^{A'}(\partial_\tau - \partial_\sigma)\psi^{A'}\} \right] \quad (11.19a)$$

$$\equiv S_{L+R} + S_R + S_L. \quad (11.19b)$$

In the above the first term written as S_{L+R} has both right and left moving components, the second term S_R of spinors has only right moving part and the third and fourth terms in curly bracket S_L have thirty-two ($A = 1, \cdots, 16$ and $A' = 1, \cdots, 16$) left moving fermion fields. We may note here that, depending on the boundary conditions imposed on the fermionic fields ψ_A and $\psi_{A'}$, we obtain two distinct types of modes, i.e. the periodic boundary condition (PBC) leads to Ramond model [9] and the antiperiodic (APBC) one to Neveu-Schwarz model [10]. For the Ramond model, one has the mode expansion,

$$\psi^A(\sigma, \tau) = \sum_n \psi_n^A exp(-2in(\tau + \sigma)), \quad (11.20)$$

with quantisation condition,

$$\left[\psi_n^A, \psi_m^B\right]_+ = \delta^{AB}\delta_{m+n,0}. \quad (11.21)$$

In the summation above since the boundary condition is periodic with period π, m and n take all integral values, positive and negative. Also, for $n = m = 0$, the algebra is merely the Dirac algebra, i.e. ψ_0^A, $A = 1, \cdots, 16$ are the Dirac matrices. The ground state here *is degenerate* with $2^8/2 = 128$ fold degeneracy corresponding to an irreducible spinor representation of $SO(16)$. For $n > 0$, ψ_n^A is annihilation operator and ψ_{-n}^A is creation operator. For the antiperiodic case, the mode expansion is

$$\psi^A(\sigma, \tau) = \sum_n \psi_n^A exp(-2in(\tau + \sigma)), \quad (11.22)$$

with summation in the above *for all half-odd-integral n* because of antiperiodic boundary condition. The quantisation condition however is

11.3. Heterotic Superstring theory

the same here as in (11.21). Here the ground state is uniquely defined. We shall now write down the mode expansion of X and S as [8],

$$X_R^i(\tau - \sigma) = \frac{1}{2}x^i + \frac{1}{2}p^i(\tau - \sigma)$$
$$+ \frac{i}{2}\sum_{n\neq 0} \frac{\alpha_n^i}{n} exp(-2in(\tau - \sigma)), \qquad (11.23a)$$

$$X_L^i(\tau + \sigma) = \frac{1}{2}x^i + \frac{1}{2}p^i(\tau + \sigma)$$
$$+ \frac{i}{2}\sum_{n\neq 0} \frac{\tilde{\alpha}_n^i}{n} exp(-2in(\tau + \sigma)), \qquad (11.23b)$$

and

$$S_R^a(\tau - \sigma) = \sum S_n^a \, exp(-2in(\tau - \sigma)). \qquad (11.23c)$$

Here x^i and p^i are the transverse coordinates and momenta in ten dimensions. Parallel to equations (11.21), we also have the quantisations

$$[x^i, p^j] = i\delta^{ij}, \qquad (11.24a)$$

$$[\alpha_m^i, \alpha_n^j] = [\tilde{\alpha}_m^i, \tilde{\alpha}_n^j] = n\delta_{m+n,0}\,\delta^{ij}, \qquad (11.24b)$$

and

$$[S_m^a, \bar{S}_n^b]_+ = (\Gamma)^{ab}\delta_{m+n,0}\,. \qquad (11.24c)$$

We then calculate the normal ordered number operator for the states. For the right handed space-time sector, the number operator becomes

$$N = \sum_{n=1}^{\infty}(\alpha_{-n}^i \alpha_n^i + \frac{1}{2}n\bar{S}_{-n}^a \Gamma^- S_n^a). \qquad (11.25)$$

Here there is no vacuum contribution, since supersymmetry is there for the right moving sector. On the other hand, for the left moving space-time sector,

$$\tilde{N}_B(X's) = \sum_{n=1}^{\infty} \tilde{\alpha}_{-n}^i \tilde{\alpha}_n^i - \frac{1}{3}. \qquad (11.26)$$

Here the contribution from the vacuum sector is $-1/3$ where we have used the ζ-function regularisation. The contribution is calculated as, $\frac{1}{2}(1 + 1 + \cdots) = \frac{1}{2}\zeta(-1) = -1/24$. Then for 8 bosonic states, total contribution is $8 \times (-1/24) = -1/3$. For the fermionic parts ψ_n^A and $\psi_n^{A'}$,

$$\tilde{N}_F(\text{Ramond}) = \sum_{n\geq 0} n\psi^A_{-n}\psi^A_n + \frac{2}{3}, \qquad (11.27)$$

and

$$\tilde{N}_F(\text{Neveu-Schwarz}) = \sum_{n\geq \frac{1}{2}} n\psi^A_{-n}\psi^A_n - \frac{1}{3}. \qquad (11.28)$$

In the Ramond sector, the contribution is of a similar type that of N_B. It is calculated as,

$$-\frac{1}{2} \times 16 \times (1+2+\cdots) = -8 \times \zeta(-1) = -8 \times -\frac{1}{12} = \frac{2}{3}.$$

Here the sign is opposite due to fermionic fields and the number of states is doubled. For the Neveu-Schwarz vector, n being a half odd integer, the calculation is,

$$-\tfrac{1}{2} \times 16 \times \left(\tfrac{1}{2}+\tfrac{3}{2}+\tfrac{5}{2}+\cdots\right)$$

$$= -8 \times \tfrac{1}{2}(\zeta(-1) - 2\zeta(-1)) = 4\zeta(-1) = 4 \times (-1/12) = -1/3.$$

We shall now identify the physical states corresponding to transverse space modes X^i's, spinorial modes S^a's, and the internal fermionic modes ψ^A_n and $\psi^{A'}_n$. The physical states will have to satisfy the condition,

$$N = \tilde{N} \qquad (11.29)$$

to maintain translational invariance in σ [8]. Let us denote $|SR(i/a)>$ as the space of right movers with bosonic "i" or fermionic "a" of the degenerate ground state, $|SL(vac)>$ as the vacuum for space left movers, $|\beta>$ as the ground state of Ramond vacuum corresponding to the 128 spinor representation of $SO(16)$, and $|NS(vac)>$ as the Neveu-Schwarz vacuum. We thus have the states labelled as,

$$|SR(i/a), SL(vac), \beta(\text{Ramond}), NS(vac)>, \qquad (11.30a)$$

which gives 128 gauge bosons and gauginos of zero mass. Here

$$N = 0, \quad \tilde{N} = -1/3 + 2/3 - 1/3 = 0$$

from equations (11.25), (11.26), (11.27) and (11.28). These 128 states in group space or internal space of a symmetry group constitute the spinor representation of $SO(16) \subset E_8$. Further the 120 gauge bosons and gauginos are obtained by operating the creation operators $\psi^A_{(-1/2)}$ $\psi^B_{(-1/2)}$ as follows. Thus the states,

11.3. Heterotic Superstring theory

$$|SR(i/a), SL(vac), [AB] >$$
$$= \psi^A_{(-1/2)} \psi^B_{(-1/2)} |SR(i/a), SL(vac), NS(vac), NS'(vac) >$$
(11.30b)

constitute the adjoint representation of $SO(16) \subset E_8$. They are obviously antisymmetric in the indices A and B, Here

$$N = 0, \quad \tilde{N} = -1/3 + (-1/3 + 1/2 + 1/2) - 1/3 = 0.$$

Thus these 128+120=248 correspond to the gauge boson and gaugino states for the E_8 group.

The second E'_8 group can be formed from the other sector of fermion fields with primed indices in a completely parallel manner.

We now consider the construction of the states for the gravity sector. These are obtained by using the creation operators $\tilde{\alpha}^j_{-1}$ on the space of right moving vacuum with indices i/a, the space left moving vacuum and the Neveu-Schwarz vacuum in both the sectors of $SO(16)$. The state may be denoted as

$$|SR(i/a), j, NS(vac); NS'(vac) >$$
$$= \tilde{\alpha}^j_{-1} |i/a, SL(vac), NS(vac), NS'(vac) > .$$
(11.30c)

Here

$$N = 0, \quad \tilde{N} = (-1/3 + 1) - 1/3 - 1/3 = 0.$$

This yields 64 spinorial modes of $N = 1$ supergravity in ten dimensions given by 56 Rarita Schwinger modes and 8 spin 1/2 modes, along with 64 bosonic modes given by 35 graviton modes, one dilaton mode, and 28 modes for the antisymmetric tensor.

The following observations about the present picture of supergravity as a low energy limit of superstring theory will be in order. Firstly we note that for all the states in equations above, the relationship (11.29) is satisfied. Further, we have subdivided the states above to be associated with gauge bosons and the gauginos of $E_8 \times E'_8$ as well as have located the gravity multiplet. However, we have not associated the expected group structure anywhere, and thus the job remains incomplete. With nontrivial analylsis, the corresponding field theory for zero mass modes with appropriate gauge and gravity properties can be established. The above analysis merely aims at a proper counting of the states to illustrate how superstring theory gets associated with supergravity in higher dimensions with supergravity being an output instead of being an input, a question which is quite relevant at a fundamental level. We may note that for the massless sector of the heterotic string, almost all the mathematical structure arises from the thirty-two left moving fermionic modes which characterise the internal symmetry group.

11.4 Discussions

The present chapter deals with supergravity in ten dimensions. Obviously, it is a very complicated theory. Further, it is a useful theory if we can compactify over six of the ten dimensions so that they are not observable.

Normally, in Kaluza Klein theories, one has a theory in higher dimensions, and then one compactifies this theory to lose space dimensions as being unobservable, and gain symmetry degrees of freedom. In the present framework we see that the symmetry group is $E_8 \times E_8'$ which is such a large symmetry group as to be unmanageable and therefore useless. We would therefore like compactification to play a totally different role here. During compactification we would like to *lose* symmetry instead of generating symmetry. We shall do this with an interplay of the structure or geometry of extra dimensions with that of the symmetry group, and study this aspect in the next chapter.

REFERENCES

1. T. Kaluza, Preuss. Akad. Wiss. Berlin, Math. Phys., K1. 966 (1921); O. Klein, Z. Phys. 37, 895 (1926).

2. E. Gliozzi, J. Scherk and D. Olive, Nucl. Phys. B122, 253 (1977).

3. E. Cremmer, B. Julia and J. Scherk, Phys. Lett 76B, 409 (1978).

4. E. Cremmer and B. Julia, Nucl. Phys. B169, 141 (1979),

5. A.H. Chamseddine, Nucl. Phys. B185, 403 (1981).

6. G.S. Chapline and N.S. Manton, Phys. Lett. 120B, 105 (1983).

7. M.B. Green and J.H. Schwarz, Phys. Lett. 149B, (1984),

8. D.J. Gross, J.A. Harvey, E. Martinec and R. Rohm, Phys. Rev. Lett. 54, 502 (1985); Nucl. Phys. B256, 253 (1985).

9. P. Ramond, Phys. Rev. D3, 2415 (1971).

10. A. Neveu and J.H. Schwarz, Nucl. Phys. B31, 86 (1971),

Chapter 12

Compactification of Higher Dimensional Theories

12.1 Introduction

In the last chapter, we have shown that the heterotic string theory in the low energy limit corresponds to the anomaly free [1] $N = 1$ supergravity with $E_8 \times E_8'$ (or $SO(32)$) gauge group in ten dimensions. But in order to have a physically realistic theory, the extra dimensions are to be compactified to an extremely small length scale so that they are unobservable at available energies. Such a compactification occurs in Kaluza-Klein type of theories, where it was in fact utilised to generate internal symmetry, i.e. the geometrical structure of the compact dimensions describes the transformations of a symmetry group. But here in ten dimensional supergravity based on heterotic string theory we already have a large symmetry group $E_8 \times E_8'$ or $SO(32)$. Hence in the present case one chooses a compactification scheme such that the gauge symmetry breaks to a lower symmetry smaller than in the original theory and thus may be physically acceptable as a grand unifying group. In the subsequent sections, we shall describe some of these schemes with the corresponding phenomenology. Some of the related topics on differential geometry are given in Appendix A, and on group theory, in Appendix B.

12.2 Compactification in 10 Dimensions

In the last chapter, we have given the $N = 1$, $d = 10$ supergravity Lagrangian [2,3]. Here the field contents are e_μ^a, $a_{\mu\nu}$, λ and ϕ corresponding to vielbein, the two form potential yielding the three form field strength, the spinor and a scalar respectively. As was shown by

Chamseddine [2], in four dimensions, one normally expects $N = 4$ supergravity. This arises as follows. Each gravitino in ten dimensions correspond to four gravitinos in four dimensions leading to $N = 4$ supergravity from merely the spinor structure and Γ-matrices in ten dimensions being expressed in terms of the same in four dimensions. In fact, a spinor in ten dimensions is equivalent to four spinors in four dimensions, and that is why we have four gravitinos. Hence in order to get rid of the extra supersymmetries one has to make a consistent choice so that three gravitinos become superheavy during compactification, but one remains light so as to leave $N = 1$ supersymmetry.

In this regard, the dynamics of compactification of six out of ten dimensions is extremely complicated and has many qualitatively different choices available. At the present stage a lot of important developments have taken place, but no positive conclusion in the context of a possible physical theory has yet been reached. For the purpose of compactification, the ten dimensional spacetime manifold is taken as a direct product of $M_4 \times K$ where M_4 is the four dimensional Minkowski space and K is a six dimensional compact space. We denote the six dimensional compact space with three complex coordinates described with indices $p = 1, 2, 3$ having the $SO(6)$ tangent space which is isometric to $SU(4)$. The Minkowski space is also described with e.g. $\mu = 0, \cdots, 3$.

With the above common features and notations regarding compactification, we shall here consider three types of approaches for the same, which, with different levels of sophistication, we regard as complementary. We shall first discuss the problem with a simple minded consideration of the generation of mass of different particles in the four dimensional space resulting from compactification. We shall then consider a simple prescription of compactification as was suggested by Witten [4]. We shall lastly discuss some topological questions regarding compactification, which are mathematically sophisticated and aesthetically as well as phenomenologically appealing [5]. Our emphasis will be mainly regarding the conceptual aspects rather than the technical details.

12.2.1 Masses in Four Dimensions

We know that the gauge symmetry gets lost when in the effective Lagrangian the vector mesons acquire a finite mass. The effective Lagrangian in four dimensions is obtained when we integrate over the variables of the compact space K, and through this process we shall show that some gauge bosons acquire a mass in four dimensions resulting in an obvious loss of symmetry. To see this, we now analyse some relevant terms in the Yang-Mills sector of $d = 10$ supergravity Lagrangian as in equation (11.18) of the last Chapter. We take here

12.2. Compactification in 10 Dimensions

some components of the gauge fields corresponding to E_8 group to have nonzero vacuum expectation value (vev) in the compact space as classical background fields. These components behave like Higgs field to obtain gauge symmetry breaking, such that effectively we have,

$$A_p(x,y) = A^B{}_p(y) + A'_p(x,y), \qquad (12.1)$$

where (x,y) are respectively the coordinates of the Minkowski space M_4 and the compact space K. $A^B{}_p(y)$ is the background field whose values depend only on K. It is necessary for it to be independent of the Minkowski space coordinates to maintain translational invariance in four dimensions. $A'_p(x,y)$ is the corresponding quantum field. We now denote the 248 gauge bosons corresponding to E_8 in the subspace $SU(3) \times E_6$ as given by

$$248 = (8,1) + (1,78) + (3,27) + (3^*, 27^*). \qquad (12.2)$$

In the above we shall take $t_{\bar{b}}$ ($\bar{b} = 1,..,8$) as the generators corresponding to (8,1), t_ω ($\omega = 1,..,78$) as the generators corresponding to (1,78), $t_{i\xi}$ ($i = 1..3, \xi = 1..27$) as the generators corresponding to (3,27), and, $t_{\bar{i}\bar{\xi}}$ as the generators corresponding to $(3^*, 27^*)$. One aims to break here the $SU(3)$ local symmetry to obtain E_6 group in the effective theory which has good phenomenological prospects for grand unification. Hence we take the background field to have a vev in the (8,1) direction, i.e. in the compact space we take

$$A^B{}_p(y) = A^{B\bar{b}}{}_p(y) t_{\bar{b}} \neq 0. \qquad (12.3)$$

We shall now analyse the Yang-Mills part of the Lagrangian from equation (11.18a) given as, with $(M, N = 0..9)$,

$$L_{\text{gauge}} = F^\theta_{MN} F^{\theta MN}. \qquad (12.4a)$$

which has three parts corresponding to Minkowski and compact spaces, i.e.

$$L^I_{\text{gauge}} = F^\theta_{\mu\nu} F^{\theta \mu\nu}, \qquad (12.4b)$$

which is the pure gauge term in four dimensions, and

$$L^{II}_{\text{gauge}} = F^\theta_{\mu p} F^{\theta \mu p}, \qquad (12.4c)$$

which gives masses to gauge bosons and

$$L^{III}_{\text{gauge}} = F^\theta_{pq} F^{\theta pq}, \qquad (12.4d)$$

which gives masses to the effective four dimensional scalars. We now use the background field vevs of equation (12.3) to calculate the gauge boson masses. This yields the mass expression (which needs appropriate normalisation) as,

$L_{\text{mass}}(\text{gauge bosons})$

$$= \tfrac{1}{4} \int d\mu(y) Tr\left\{ [A'_\mu(x), A^B{}_p(y)][A'^\mu(x), A^{Bp}(y)]\right\}. \qquad (12.5a)$$

In the above, $d\mu(y)$ is the measure in the compact space K, and we have retained only the first term in the expansion of $A'_\mu(x,y)$ of equation (12.1) as a function of y. We have also used corresponding to equation (12.2) the implicit notation

$$A_M = A_M{}^{\bar{b}} t_{\bar{b}} + A_M{}^\omega t_\omega + \cdots \qquad (12.5b)$$

with the terms corresponding to $(3,27)$ and $(3^*,27^*)$ representations not being written. What is relevant is that from equation (12.5) only the gauge bosons $A'^\omega{}_\mu(x)$ will remain massless as the generators of $(8,1)$ commute with those of $(1,78)$, whereas the gauge bosons of E_8 corresponding to the other representations on the right hand side of equation (12.2) will generally become massive as the corresponding commutators do not vanish. Thus the surviving symmetry group here becomes E_6 after compactification.

We shall next obtain the massless sector for matter that will appear in the effective Lagrangian. The fermionic mass term of the gauge sector in the Lagrangian as in equation (11.18b) is generated by,

$$L_{\text{gaugino}} = \bar{\chi} \Gamma^M D_M \chi + \cdots. \qquad (12.6)$$

In the above, χ is a spinor in 10 dimensional space, which we can specify with two indices (a,α) with $\alpha = 1,2$ describing the spinorial structure for $SO(1,3)$ and $a = 1,\cdots,4$ describing the "balance" of spinor indices as needed for the description of spinors in the ten dimensional space [2]. This decomposition corresponds to the fact that one spinor in ten dimension corresponds to four spinors in four dimensions. Further, when $M = p$, an index in the compact space, the covariant derivative D_p in equation (12.6) is given as

$$D_p = \partial_p + A_p + \omega_p \qquad (12.7)$$

where A_p is the gauge connection and ω_p is the spin connection for the assumed nontrivial geometry of the compact space. We may now consider for example the gauginos corresponding to the $(3^*,27^*)$ representation. Further, we take $SU(3)$ holonomy for the compact space K. The complex coordinates defined in the beginning were really chosen with this in mind. Then the spinor indices decompose to

12.2. Compactification in 10 Dimensions

$$4 = 3 + 1 \qquad (12.8)$$

i.e. the four spinors in four dimensions decompose to a triplet of spinors of the $SU(3)$ holonomy group and a singlet for the same group. We now consider the spinors for $a = 1,2,3$ for the triplet as above, and use the decomposition

$$\chi_{a\alpha}^{\bar{i}\bar{\xi}} = \delta_{ia}\chi_{\alpha}^{\bar{\xi}} + (t_{\bar{b}})_{ia}\chi_{\alpha}^{\bar{b}\bar{\xi}} \qquad (12.9)$$

where $\chi^{\bar{\xi}}$ is a 27^*-plet spinor in Minkowski space with E_6 symmetry. We note that the mass term for the gauginos arises from the integral

$$\int d\mu(y)\bar{\chi}\Gamma^P D_P \chi$$

and thus if spin connection equals gauge connection, then the 27^*-plet spinors in equation (12.7) remain massless. We note that $R \equiv SU(3) \subset E_8$ and $\tilde{R} \equiv SU(3) \subset SO(6)$ are the two groups involved here, the second one being the holonomy group, and the masslessness of the 27^*-plet, or in a similar way, of the 27-plet, arises from the $R + \tilde{R}$ invariance with an interlinking of the group space with the geometry of the compact space. his interplay of the space structure (spin connection) with the group structure (gauge connection) is a special feature during compactification which results in symmetry breaking and gives rise to E_6 gauge group for the residual symmetry. The massless "matter" sector consists of the 27 and 27^* representations of this residual E_6 gauge group.

We next note that the single gravitino in ten dimensions leads to four gravitinos in four dimensions. Let us now examine the gravitino masses to see that in four dimensions three out of the four gravitinos become massive, leaving one as massless, so that only $N = 1$ supersymmetry remains unbroken. We note that from equation (11.18b) the mass term for the gravitino in four dimensions arises from integrals like

$$\int \bar{\psi}_\mu \sigma^{\mu\nu} \Gamma^P D_P \psi_\nu d\mu(y). \qquad (12.10)$$

The gravitino does not carry any gauge structure, and hence the gauge connection for the covariant derivative D_p as in equation (12.7) here vanishes. But, corresponding to the decomposition (12.8), the spin structure in compact space for the triplet of gravitinos does not vanish, and thus the corresponding three gravitinos become massive. In the same decomposition, the spin structure for the singlet vanishes, and the corresponding gravitino remains massless, which yields $N = 1$ residual supersymmetry. We note that in equation (11.8b) we have assumed that for the compact space we can consistently take $H_{pqr} = 0$, as was considered in Ref.[5]. Also, we have explicitly considered the generation of masses during compactification to see which particles remain massless, as well as to see how compactification generates masses for a large number of particles. These masses naturally will be in the same scale as the compactification scale. When we eliminate these heavy particles, we shall naturally have "four fermion" type of interactions, which however will be suppressed by the compactification scale, expected to be of the same order as the Planck scale, and thus will not be observable.

12.2.2 Witten's prescription

We shall now truncate the ten dimensional supergravity Lagrangian as of last chapter to obtain a four dimensional supergravity Lagrangian in four dimensions as was considered in chapter 9 through a few prescriptions [4].

Let us now first consider the gravity sector. As before we shall assume $SU(3)$ holonomy for the compact space K. This sector is a singlet in gauge space. Further, we shall assume that only $SU(3)$ singlets survive after compactification to four dimensions, the remaining fields becoming superheavy. For this purpose, the complex coordinates already taken will be quite convenient (see Appendix). For the metric we shall write

$$g_{pq^*} = \delta_{pq^*} exp(\sigma), \qquad (12.11a)$$

where σ is a real scalar field and is a singlet under the holonomy group, and thus survives in four dimensions. In terms of the conventional real coordinates I, J, for the compact space K. the above equation really means that

$$g_{IJ} = \delta_{IJ} exp(\sigma). \qquad (12.11b)$$

For the components of the antisymmetric potential in the compact space K, we also write

$$a_{p\bar{q}} = \frac{i}{2}(\delta_{p\bar{q}} - \delta_{q\bar{p}})a, \qquad (12.12)$$

which, in terms of compact space indices $4,\cdots,9$ leads to $a_{45} = a_{67} = a_{89} = a = -a_{54} = -a_{76} = -a_{98}$, and that all other components vanish. The fact that a as above is a singlet under $SU(3)$ holonomy is transparent with the use of complex coordinates when we recognise that the singlet corresponds to the singlet in $3 \times 3^* = 1 + 8$. Further, with only the indices of the four dimensional space, we substitute the field strength

$$H_{\mu\nu\rho} = \epsilon_{\mu\nu\rho\sigma}\, \partial^\sigma \lambda$$

12.2. Compactification in 10 Dimensions

where λ is an $SU(3)$ holonomy singlet pseudoscalar field. We also note that components of the antisymmetric potential or the metric with the index structure of one being in the Minkowski space and the other being in the compact space can none be singlets under $SU(3)$ holonomy, and thus as per the present prescription, none of them will survive in four dimensions. Thus during compactification the gravity sector of ten dimensions has, besides the gravity sector of four dimensions, only three extra fields in four dimensions given as σ, a, and λ.

We next consider the gauge nonsinglet fields of the ten dimensional supergravity Lagrangian. As before, we have $R = SU(3)$ is the holonomy group of K and $\tilde{R} = SU(3) \subset E_8$ is a subgroup of the gauge group, and the prescription for the fields in four dimensions after compactification will be that they must be invariant under combined R and $\tilde{R}$ transformations. We now note that the 248×6 gauge vector components of the extra dimensions are scalars in four dimensions, and are given as

$$A_p^{8,1}, \ A_p^{1,78}, \ A_p^{3,27}, \ A_p^{3^*,27^*}.$$

Amongst these only 27 and 27^* have combined R and $\tilde{R}$ invariance. The identification is parallel to equation (12.9) and is given through

$$A_p^{\bar{i},\bar{\xi}} = \delta_p^i C^{\bar{\xi}} \qquad (12.13)$$

for a 27^*-plet of scalars, and a similar result for the representation $(3,27)$. The field content in four dimensions after compactification thus becomes highly restricted due to the prescription of R and $\tilde{R}$ invariance, the possible origin of which as a dynamical question is unclear.

We can then simplify the Lagrangian in four dimensions as [4]

$$\begin{aligned}L = &-\frac{1}{2}R^{(4)} - 3\partial_\mu\sigma\partial^\mu\sigma - \frac{9}{16}\frac{\partial_\mu\phi\partial^\mu\phi}{\phi^2} \\ &- \frac{3}{2}e^{-2\sigma}\phi^{-3/2}\left(\partial_\mu a - i\frac{\kappa}{\sqrt{2}}C^{\bar{\xi}}D_\mu C^{\xi}\right)^2 \\ &- \frac{3}{4}\phi^{-3/2}e^{6\sigma}H^{\mu\nu\rho}H_{\mu\nu\rho} - \frac{1}{4}\phi^{-3/4}e^{3\sigma}Tr[F_{\mu\nu}F^{\mu\nu}] \\ &- 3e^{-\sigma}\phi^{-3/4}(D_\mu C^{\xi})^*(D^\mu C^{\xi}) - \frac{8}{3}g^2\phi^{-3/4}e^{-5\sigma} \times \left|\frac{\partial W}{\partial C^{\xi}}\right|^2 \\ &- \frac{9}{2}g^2\phi^{-3/4}e^{-5\sigma}\sum_i (C^*t_iC)^2 \\ &- 8\kappa^2 g^2\phi^{-3/2}e^{-6\sigma}|W|^2, \end{aligned} \qquad (12.14)$$

where

$$W = 8\sqrt{2}g \, d_{\xi\eta\zeta} \, C^\xi C^\eta C^\zeta \qquad (12.15)$$

is a trilinear invariant from the three 27-plet fields and $H_{\mu\nu\rho}$ contains the Chern Simons term as in Ref [6] as mentioned in the last chapter. We note that equation (12.14) is still not in the form of a supergravity Lagrangian. To do this we recall that a supergravity Lagrangian is completely defined when we identify the function G as in equation (9.12). This function in turn is defined through the superpotential $g(S)$ and the function ϕ as from equations (9.6) and (9.7). We may however replace ϕ by the Kahler potential J as in equation (9.16). We shall now define W as the super potential, and construct a Kahler potential as follows. With three antisymmetric indices, we define a pseuedoscalar field D through the equation

$$\phi^{-3/2} e^{6\sigma} H_{\mu\nu\rho} = \epsilon_{\mu\nu\rho\sigma} \, \partial^\sigma D \qquad (12.16)$$

and then define two fields S and T through the equations

$$S = e^{3\sigma} \phi^{-3/4} + 3i\sqrt{2}D \qquad (12.17a)$$

and

$$T = e^\sigma \phi^{3/4} - i\sqrt{2}a + C^{\bar{\xi}} C^\xi. \qquad (12.17b)$$

With the above fields corresponding to chiral fields, we consider next the Kahler potential given as

$$J = -ln(S + S^*) - 3 \, ln(T + T^* - 2C^*C). \qquad (12.18)$$

With the superpotential as in equation (12.15) and the Kahler potential as in equation (12.18), we can then show that the corresponding supergravity Lagrangian as derived in Chapter 9 becomes the same as in equation (12.14). This way, effectively through the assumption that $SU(3)$ holonomy of the compact space K couples to the $SU(3)$ subgroup of E_8, one obtains the four dimensional supergravity Lagrangian with $N = 1$ supersymmetry as above when it is assumed that in four dimensions only fields which are singlets under the above combined transformations survive below the compactification scale. We may also note that the Lagrangian of equation (12.14) has a classical scale invariance which can only disappear when σ or ϕ fields attain a vacuum expectation value.

12.2.3 Geometry of extra dimensions

As stated earlier, we wish to have symmetry breaking with compactification. We shall see here how it depends on the geometry of the

12.2. Compactification in 10 Dimensions

extra dimesnions. For this purpose clearly some of the fields which are scalars in four dimensions should have nonzero values in the compact space of extra dimensions. However, no fermion can have nonzero value even in extra dimensions, since fermions in ten dimensions always decompose to fermions in four dimensions and naturally transform under a Lorentz transformation. However, under a supersymmetry transformation with fermionic increments, a nonzero bosonic field will generally give a nonzero increment to a fermionic field. We have to ensure that the bosonic fields in compact space are so constrained that the fermionic increments due to supersymmetry vanish. This is a consistency requirement for Lorentz invariance after compactification. We shall now proceed to illustrate this [5], where we note the final results instead of the details of the sophisticated algebra. We wish to mention that the supersymmetry transformation under consideration will be in four dimensions, since we wish to have $N = 1$ supersymmetry in four dimensions after compactification and thus will refer to a restricted class of supersymmetry transformations in ten dimensions.

We note that since the fermionic fields vanish for the vacuum, the increments under a supersymmetry transformation for Bose fields automatically vanishes. In order to consider the increments of the fermion fields, it will be useful to explicitly write down the Γ matrices in ten dimensions in terms of those in four dimensions. As stated we shall briefly present here the concepts of Ref [5] and therefore shall choose the same notations. We shall thus take

$$[\Gamma^M, \Gamma^N] = -2\eta^{MN}. \tag{12.19}$$

Here $\eta^{MN} = diag(1, -1, -1, \cdots, -1)$ in ten dimesnions. We take the Majorana representation so that Γ^0 is real and antihermitian, and the remaining Γ^M-matrices are real and hermitian. We can then take [6]

$$\Gamma^\mu = \gamma^\mu \otimes I; \quad \Gamma^m = \gamma_5 \otimes \gamma^m, \tag{12.20a}$$

with

$$\gamma_5 = \frac{i}{4!}\epsilon_{\mu\nu\rho\sigma}\gamma^{\mu\nu\rho\sigma}. \tag{12.20b}$$

Clearly in the above γ^μ's are Dirac matrices in Minkowski space and γ^m's are the same for the six dimensional internal space. In the internal space we also take the parallel of γ_5 as

$$\gamma = \frac{i}{6!}g^{1/2}\epsilon_{mnpqrs}\gamma^{mnpqrs}. \tag{12.21}$$

γ^0 is antihermitian, the remaining γ^μ's are real and hermitian, and, γ_5, γ, and γ^m's are imaginary and hermitian. As earlier such an index structure will help us to separate the dependance of the spinor indices for the internal space and the Minkowski space.

We next consider the increments for the fermionic fields. We recall that here we have to consider a spin 3/2 field ψ_M, the spin 1/2 field λ, and the gaugino spin 1/2 fields χ^a with a as the group index. The boson fields consist of the metric g_{MN}, the two form potential B_{MN} which gives rise to the three form field strength H_{mnp} including the Yang Mills three form ω_{3Y} and the Lorentz three form ω_{3L} as in equations (11.14) and (11.16), a scalar field ϕ, and the Yang Mills field strengths $F^a{}_{MN}$. With the above we now note the increments of the spinor fields. With the supersymmetry increment being ϵ, we then have the following equations for the vanishing of the spinorial increments:

$$\delta\psi_\mu = D_\mu \epsilon + \frac{\sqrt{2}}{32} e^{2\phi}(\gamma_\mu \gamma^5 H)\epsilon = 0, \qquad (12.22a)$$

$$\delta\psi_m = D_m \epsilon + \frac{\sqrt{2}}{32} e^{2\phi}\left[\gamma_m H + 12 H_m\right]\epsilon = 0, \qquad (12.22b)$$

$$\delta\lambda = \sqrt{2}(\gamma^m D_m \phi)\epsilon + \frac{1}{8} e^{2\phi} H \epsilon = 0, \qquad (12.22c)$$

and

$$\delta\chi^a = -\frac{1}{4} e^\phi F^a{}_{mn} \gamma^{mn} \epsilon = 0. \qquad (12.22d)$$

The first two equations above give the changes of the gravitino field respectively in Minkowski space and in the compact space, and, the third corresponds to the spin 1/2 field of the gravity sector, and the fourth to that of the gaugino field. We also note that we have substituted

$$H = H_{pqr}\gamma^{pqr}; \quad H_m = H_{mpq}\gamma^{pq}. \qquad (12.23)$$

γ^{pq} and γ^{pqr} are appropriately normalised antisymmetric products of the γ-matrices of the compact space. All the increments of equation (12.22) as stated vanish so that Lorentz invariance in four dimensions is maintained.

We want that after compactification only $N = 1$ supersymmetry should survive. We want to know what should be the nature of the compact space so that this happens. We note that in fact equations (12.22) are eigenvalue equations for ϵ with zero eigenvalue. In addition to the above equations, we shall make one more simplifying assumption, which is that the Minkowski space M_4 is maximally symmetric. This is equivalent to

12.2. Compactification in 10 Dimensions

$$R_{\mu\nu\rho\sigma} = \kappa(g_{\mu\rho}g_{\nu\sigma} - g_{\mu\sigma}g_{\nu\rho}), \qquad (12.24)$$

with the parameter κ being posititve for de Sitter space, negative for anti-de Sitter space, and, zero for Minkowski space. We now note that from equation (12.22a)

$$[\nabla_\mu, \nabla_\nu]\epsilon - \left(\frac{1}{16}\right)^2 e^{4\phi}(\gamma_{\mu\nu} \otimes H^2)\epsilon = 0. \qquad (12.25)$$

Further,

$$[\nabla_\mu, \nabla_\nu]\epsilon = \frac{1}{4} R_{\mu\nu\rho\sigma} \gamma^{\rho\sigma}\epsilon = \frac{\kappa}{2}\gamma_{\mu\nu}\epsilon, \qquad (12.26)$$

which yields that

$$\gamma_{\mu\nu} \otimes \left(\frac{\kappa}{2} - \left(\frac{1}{16}\right)^2 e^{4\phi} H^2\right)\epsilon = 0. \qquad (12.27)$$

Since antisymmetric $\gamma_{\mu\nu}$ is invertible, we thus obtain that

$$\left(\frac{\kappa}{2} - \left(\frac{1}{16}\right)^2 e^{4\phi} H^2\right)\epsilon = 0. \qquad (12.28)$$

Next, using equation (12.22c), we can prove that

$$(\nabla_m \phi)(\nabla^m \phi) = -\kappa. \qquad (12.29)$$

Since the internal manifold K is compact, ϕ must have an extremum at some place, which shows that we must have $\kappa = 0$. Hence the Minkowski space M_4 must be flat. It may be noted that the flatness of the Minkowski space has been an output from the above when we merely assumed that it is maximally symmetric as in equation (12.24). Equation (12.28) now yields that

$$H\epsilon = 0. \qquad (12.30)$$

Also, equation (12.22b) reduces to

$$\nabla_m \epsilon \equiv \left(\nabla_m - \frac{3\sqrt{2}}{8} e^{2\phi} H_m\right)\epsilon = 0. \qquad (12.31)$$

In (12.32) we have regarded ∇_m as the covariant derivative with an additional contribution from torsion proportional to H_{pqr}. Equations (12.30) and (12.31) influence the structure of the compact manifold K. We shall later construct three dimensional complex space K corresponding to the six real dimensions of the internal space with the construction that the torsion H_{pqr} is zero, which will imply that ϵ is covariantly constant with ∇_m giving the covariant derivative; for the present however we retain the above general structure, and show that the space of the extra dimensions is a Kahler manifold.

From equations (12.30) and (12.31) we obtain that

$$\gamma^m \nabla_m \epsilon = 0. \tag{12.32}$$

Thus the compact space K admits a covariantly constant spinor. Further this fermionic parameter for supersymmetric transformations when subjected to Weyl condition

$$(\gamma_5 \otimes \gamma)\epsilon = -\epsilon, \tag{12.33}$$

can be expanded as

$$\epsilon = \sum \xi_\alpha \otimes \eta_\alpha, \tag{12.34}$$

where ξ_α are four spinors with real fermionic components, and η_α are four spinors with real bosonic components. They are not eigenspinors of γ_5 and γ separately. We thus assume that there exist at least two real spinors η with bosonic components in the internal space. We choose the normalisation $\eta^\dagger \eta = 1$, and then construct, as in equation (A.19) of Appendix A, the almost complex structure

$$J_m{}^n = -i\eta^\dagger \gamma_m{}^n \eta, \tag{12.35a}$$

satisfying the equation

$$J_m{}^p J_p{}^n = -\delta_m{}^n, \tag{12.35b}$$

which is the same as equation (A.3b). We may further verify that the Niejenhuis tensor as in equation (A.3c) vanishes. Hence K is a complex hermitian manifold.

We shall now look for suitable choices for the internal manifold K. For this puropose, let us make the simplifying assumption that $H_{pqr} = 0$. We then have, from equation (12.31),

$$\nabla_m \eta = 0. \tag{12.36}$$

We can then show that $R_{mn} = 0$, i.e. the manifold K is Ricci flat. Hence, we shall take the candidate manifold for the internal space as a Calabi Yau space, discussed in Appendix A. Such a space has $SU(3)$ holonomy; i.e. the tangent vectors of the space decompose to multiplets of $SU(3) \subset SO(6)$ in the six dimensional manifold K and admits chiral fermions. Because the fermions in ten dimensions obey Weyl condition, fermions which are left or right handed in four dimensions are left or right handed on K.

Such spaces must have vanishing first Chern class. We have discussed the construction of such spaces in Appendix A. One of the simplest examples [5] is the CP^4 space with one polynomial condition as discussed in subsection A.10.1 of the appendix. Let CP^4 be described by five homogeneous coordinates z_i ($i = 1, .., 5$), and we take the polynomial condition as

12.2. Compactification in 10 Dimensions

$$\sum z_i^5 = 0 \qquad (12.37)$$

as in equation (A.31). The Euler number of this manifold K_0 from equation (A.47) is -200. Hence, as mentioned in the appendix, it will have one hundred generations of quark and leptons, a situation which is not acceptable. This is changed when we introduce an equivalence relation with a discrete symmetry $Z_5 \times Z_5 = G$ freely acting on the manifold K_0 as stated in equations (A.32) and (A.32') with the first Z_5 being a suitable reordering of homogeneous coordinates, and the second Z_5 generated through a fifth root of unity. Then the Euler number of K_0/G becomes -8, and the number of generations, four. As per the present collider experiments for the number of light neutrinos (which is three), four generations is probably still too large.

Another model constructed by Tian and Yau, starting from the $CP^3 \times CP^3$ space with three polynomial conditions and the discrete symmetry Z_3 has three generations for quarks and leptons, which is better. This has been discussed in subsections 10.2 and 11.3 of Appendix A. There are a few more examples which have been subsequently constructed. We shall however mostly confine our attention to the above two to discuss the techniques and the concepts.

We can expand spinors on K corresponding to $SU(3)$ holonomy as in equation (A.22). When the space is multiply connected, we may expect that the covariantly constant spinor ζ which is an $SU(3)$ holonomy singlet might have a phase factor as one goes round a closed curve which is not contractible. It was shown in Ref. [5] that this phase shall not be there.

Let us examine the gauge field configurations in K which will permit the existence on $N = 1$ unbroken supersymmetry. With complex indices p, q describing K, it has been shown Ref. [5] that we shall have

$$F_{pq} = F_{\bar{p}\bar{q}} = 0, \quad g^{p\bar{q}} F_{p\bar{q}} = 0. \qquad (12.38)$$

Consistent with the above, if we take gauge connection is equal to spin connection, $N = 1$ supersymmetry will be maintained.

We note that the expression (11.14) corresponds to the fundamental representation of $SO(32)$. On the other hand trace of generators in group space is dependant on the representation. For anomaly cancellation, trace is needed in the calculation of loops corresponding to Ω_{12} of equation (11.17). Taking this into account, for the adjoint representation of $SO(32)$ or for $E_8 \times E_8'$ we have in fact [5]

$$\omega_{3Y} = \frac{1}{30} Tr\left[A \wedge F - \frac{1}{3} A \wedge A \wedge A\right] \qquad (12.39)$$

explicitly instead of (11.14), since the Lorentz part of the calculation in (11.16) does not change. We now consider the decomposition of the

248 representation of E_8 to $SU(3) \times E_6$ as in equation (12.2). Since we can write the generators in octet as $t_a \times I - I \times t_a$ with t_a being a generator for the representation 3 of $SU(3) \subset E_8$, it is clear that for the octet the contribution to the second order Casimir invariant is three times that of $(3+3^*)$. Further, for $(3,27)$ and $(3^*,27^*)$ we have again a factor of twenty-seven. Since in the space K the Lorentz anomaly in equation (11.16) will correspond to the vector representation $(3+3^*)$ for the application of $SU(3)$ holonomy, the factor thirty will cancel with $(3+27)$ as shown in equation (12.39) above.

We further take that $SU(3)$ holonomy in K equals negative of background gauge field for the $SU(3)$ subgroup of the gauge group $E_8 \supset SU(3) \times E_6$ in the internal space K. This is equivalent to the $R+\tilde{R}$ invariance introduced in the last subsection. Since gravitinos are gauge singlets, from equation (12.10) we obtain the $SU(3)$-holonomy singlet gravitino as massless and the others as massive. Also, b_{11} copies of 27 and b_{21} copies of 27^* to be massless quarks and leptons will be there which depends on the structure of the internal space as discussed above and in Appendix A. Here b_{pq} are the Betti-Hodge numbers of K, explained in appendix A.

There are further ramifications when the space is multiply connected as considered above. If g be an element of the discrete group, and y be a point on the manifold, in the presence of discrete symmetry we have $gy = y$. Then we may consider

$$T(g) = exp\Big\{ i\tilde{g} \int_y^{gy} A_m dy^m \Big\}. \qquad (12.40)$$

where the integral as above is over a closed contour in K, but need not vanish even when the field strength vanishes. The nonvanishing of the above integral can also give rise to symmetry breaking, and will be considered in the next section. This is the mechanism of symmetry breaking through Wilson loops and will be considered in some detail later. Since the field strengths vanish, at the tree level, the energy here is the same as that of perturbative vacuum and thus symmetry broken phase becomes permissible.

12.3 Superstring based supergravity

When we consider superstring based supergravity in four dimensions, we really examine the $N = 1$ supergravity resulting from ten dimensional supergravity as described in the previous section. The infinity of modes corresponding to string theory present in the Planck scale are ignored. We shall proceed with the phenomenological applications in that context.

12.3. Superstring based supergravity

We note that the theoretical considerations as in the last subsection on compactification had the following broad features. Firstly, we noted that the spin connection of extra dimensions may be the same as the gauge connection for the background gauge fields which will thus have nonvanishing components corresponding to the extra dimensions. We have seen that this gives rise to $N=1$ supersymmetry along with symmetry breaking of E_8 in ten dimensions to E_6 in four dimensions. Secondly, to obtain a reasonably small number of generations of quarks and leptons, we needed to have the manifold K corresponding to the extra dimensions as multiply connected. Hence nonvanishing of expressions like those in (12.40) will give rise to further symmetry breaking. This will in particular break E_6 symmetry to smaller semisimple groups. We now proceed to consider this mechanism in greater detail.

12.3.1 Wilson loops

Symmetry breaking through Wilson loops, or the flux breaking mechanism, or the Hosotani mechanism [7], as stated above take place through nonvanishing of closed loop integrals in a multiply connected internal space. Here the field strengths vanish everywhere, but the potentials are nontrivial. Let us first give an example of the same with the discrete group $Z_2 \times Z_2 \times Z_2 \equiv G$ with the internal space K_0 of dimension six being described by six polar angles $(\theta_i, \phi_i : i = 1, 2, 3)$. Z_2' as above are generated through transformations g which take a point to antipodal points, i.e. $g(\theta_i, \phi_i) = (\pi - \theta_i, \phi_i + \pi)$. Then the compact space $K = K_0/G$ becomes

$$(S^2/Z_2) \times (S^2/Z_2) \times (S^2/Z_2). \tag{12.41}$$

We now assume that through some mechanism we already have $E_6 \to SU(3)_C \times SU(3)_L \times SU(3)_R \equiv G_{333}$ and then we consider the flux breaking of G_{333} to a smaller group in order to illustrate the ideas better. We wish to obtain that $SU(3)_C$ remains unbroken, $SU(3)_L$ breaks to $SU(2)_L \times U(1)_Y$ and that $SU(3)_R$ breaks to $U(1)_{3R} \times U(1)_{8R}$. To achieve this, let us take the polar components of the gauge fields as nonzero constants as given below.

$$A^{8L}_{\phi_1} = v_{8L}, \quad A^{8R}_{\phi_2} = v_{8R},$$

and

$$A^{8R}_{\phi_3} = v'_{8R}, \quad A^{3R}_{\phi_3} = v_{3R}. \tag{12.42}$$

We now integrate over K with $\theta_i = \pi/2$ and taking $0 < \phi_i < \pi$ respectively which define closed contours on K corresponding to action of g_i on K_0. Further, we take the corresponding line elements for the integrals for each ϕ_i as given by $dy_i = m_i^{-1} d\phi_i$ where m_i^{-1} are length scales associated with compactification dimensions. We also take the group elements as defined in equation (12.40) to satisfy the following specific conditions:

$$T(g_1) \equiv exp\left\{i\tilde{g}t_{8L}\int v_{8L}(d\phi_1/m_1)\right\}$$

$$= \begin{pmatrix} -1 & 0 & 0 \\ 0 & -1 & 0 \\ 0 & 0 & 1 \end{pmatrix}_L, \qquad (12.43a)$$

$$T(g_2) \equiv exp\left\{i\tilde{g}t_{8R}\int v_{8R}(d\phi_2/m_2)\right\}$$

$$= \begin{pmatrix} -1 & 0 & 0 \\ 0 & -1 & 0 \\ 0 & 0 & 1 \end{pmatrix}_R, \qquad (12.43b)$$

and

$$T(g_3) \equiv exp\left\{i\tilde{g}(d\phi_3/m_3)\left[t_{3R}\int v_{3R} + t_{8R}\int v'_{8R}\right]\right\}$$

$$= \begin{pmatrix} -1 & 0 & 0 \\ 0 & 1 & 0 \\ 0 & 0 & -1 \end{pmatrix}_R. \qquad (12.43c)$$

In the above, $\tilde{g}$ is the gauge coupling constant. It is clear that $T(g_1)$ generates a $\tilde{Z}_2$ in $SU(3)_L$ and that $T(g_2)$ and $T(g_3)$ generate $\tilde{Z}_2 \times \tilde{Z}_2$ in $SU(3)_R$ such that the above gives a map of G on K_0 defining multiple connectivity of K to $\tilde{G}$, which is a discrete subgroup of G_{333} or E_6 isomorphic to G. Obviously it is very similar to the equality of connection in internal space and gauge connection, and again here we have found a link of multiply connectedness of internal space with the gauge group.

It is obvious that equations (12.43) will be true if we have $v_{8L} = 2\sqrt{3}m_1/\tilde{g}$, $v_{8R} = 2\sqrt{3}m_2/\tilde{g}$, $v_{3R} = m_3/\tilde{g}$ and $v'_{8R} = m_3\sqrt{3}/\tilde{g}$ as can be derived from the two diagonal Gellmann generators t_3 and t_8 of $SU(3)$. Thus the vacuum expectation values are of the order of compactification scale, and they are quantised. The energy density due to this compactification however is the same as the usual vacuum energy density. The above vacuum expectation values thus are determined not through the minimisation of energy but through flux breaking. When

12.3. Superstring based supergravity

we include quantum corrections, the vacuum configuration with flux breaking will have different energy than vacuum without it. This calculation is a difficult task and has been possible to work out only in very simple cases [8]. E.g. it needs a generalisation of equations like (12.42) to obtain an expansion with all modes. In the present case they are spherical harmonics with suitable conditions and are simple. But in more complicated cases say for Calabi Yau spaces, these are too difficult to construct and utilise in any calculation.

The above model was constructed only as a simple illustration of the flux breaking mechanism. In fact, we note that the G_{333} content of the adjoint representation of E_6 and of 27-plet are given by

$$78 = (8,1,1) + (1,8,1) + (1,1,8) + (3,\bar{3},\bar{3}) + (\bar{3},3,3), \qquad (12.44a)$$

and

$$27 = (3,3,1) + (\bar{3},1,3) + (1,\bar{3},\bar{3}). \qquad (12.44b)$$

From this we can easily see that the quantum numbers of quarks and leptons are not correctly given in G_{321} when symmetry breaking takes place, and thus the model is unphysical. We shall now proceed to construct other examples starting from Calabi Yau spaces having a greater chance of representing the physical world.

Before we discuss further examples, the following observations, now overdue, shall be relevant. On a manifold K where gauge transformations are defined, and we have a field $\psi(y)$ which is a gauge multiplet, then, for a curve γ connecting the points y and y', the gauge covariant parallel displacement is given as

$$\psi(y') = P \, exp\{i\tilde{g} \int_\gamma A_m d\eta^m\} \psi(y), \qquad (12.45)$$

where P stands for path ordering [9]. However, if $y' = gy$ with $K = K_0/G$ and $g \epsilon G$, then γ is a closed curve in K. Hence for any "acceptable" gauge multiplet on K we shall have

$$\psi(y) = P \, exp\{i\tilde{g} \int_\gamma A_m d\eta^m\} \psi(y) \qquad (12.46a)$$

$$\equiv T(g)\psi(y). \qquad (12.46b)$$

Since here we are taking background classical fields and diagonal generators while discussing symmetry breaking using the adjoint representation, path ordering is immaterial, and $T(g)$ of equation (12.46b) becomes the same as that in equation (12.40) considered earlier. Group theoretic constraints arising from this so that the symmetry breaking pattern contains the standard group G_{321} are considered in the the next subsection and further in appendix, subsection B.5.1. In the literature, these have been considered from different points of view by many authors [10-23].

12.3.2 Spaces with $Z_5 \times Z_5$ discrete symmetry

Parallel to equations (12.40) or (12.46) above, let us consider the group element of E_6 generated through the Wilson loop as, now absorbing the gauge coupling in the vector field,

$$T(g) = exp\{it_a \int_y^{gy} A^a{}_m(y) d\eta^m\}. \qquad (12.47)$$

As before $g \epsilon G = Z_5 \times Z_5$ and t_a's are generators of the Cartan subalgebra of E_6. As discussed in the last subsection and Appendix B, this will give rise to a symmetry breaking axis in E_6. In the dual basis [10], we may take this generally as $\bar{c} = (a, b, c, d, e, f)$. With a particular embedding of the standard group G_{321} in E_6, when we want G_{321} to remain unbroken, the symmetry breaking axis gets restricted as has been obtained in Appendix B. In fact we have

$$\bar{c} = (a, -a, c, d, -a, 0), \qquad (12.48)$$

which is the same as equation (B.19) as derived in Appendix B. We shall now examine the different values of a, c and d which will be generated through the T's of equation (12.47) corresponding to $Z_5 \times Z_5$ discrete symmetry.

With α and β as nontrivial fifth roots of unity, let the $T(g)$'s of equation (12.47) be generated by the matrices

$$\begin{pmatrix} 1 & 0 & 0 \\ 0 & 1 & 0 \\ 0 & 0 & 1 \end{pmatrix}_C \times \begin{pmatrix} \alpha & 0 & 0 \\ 0 & \alpha & 0 \\ 0 & 0 & \alpha^3 \end{pmatrix}_L \times \begin{pmatrix} \beta & 0 & 0 \\ 0 & \beta^4 & 0 \\ 0 & 0 & 1 \end{pmatrix}_R. \qquad (12.49)$$

We thus obtain the group defined by the maps G on K_0 to as corresponding to $\tilde{G}$, a subgroup of E_6, here through the subgroup G_{333} of E_6 exactly as in the last example. Equation (12.49) corresponds to the identification of the Dynkin labels in dual basis of equation (12.48) as $\alpha = exp(2\pi ic/5)$, $\beta = exp(2\pi id/5)$ and $\alpha\beta = exp(-2\pi ia/5)$, as can be derived from equation (B.7) in appendix for the eigenvalues of diagonal generators. Thus here a, c and d are not independent. Further they have the equivalence relations as $m = n \mod(5)$ corresponding to the fifth roots of unity for the group Z_5 [13].

In order to apply these techniques, it is useful to write half of the nonnegative roots of E_6. We give them in Table 1 of appendix B along with an identification of the simple roots in terms of the subgroups of E_6.

12.4 Three Generation Models

We shall now consider a class of three generation superstring based models with the ideas and technology as mentioned above. The models we shall discuss here are constructed with Tian Yau manifold introduced in subsections 10.2 and 11.3 of Appendix A as the six dimensional internal space which is compactified. We discuss this in some detail to illustrate the methods as well as because phenomenologically they are most viable.

Before getting into details, it is worthwhile to note the following broad features of the models. In continuation with what is discussed earlier, the internal manifold which is compactified is a Calabi Yau space. It is multiply connected with Z_3 as the discrete symmetry group. With Wilson loop mechanism of symmetry breaking, the residual symmetry group becomes a subgroup of E_6. These symmetry groups could be $SU(6) \times U(1) \equiv G_{61}$ or $SU(3)_C \times SU(3)_L \times SU(3)_R \equiv G_{333}$. The nature of the compact space generates multiple copies of leptons as well as quarks corresponding to the residual symmetry. Clearly the above two symmetries arising from compactification can only be intermediate symmetries. The most remarkable feature of the models is that the Yukawa couplings can be calculated from the geometry of the internal manifold except for an overall constant. Using the expected or calculated forms of Yukawa couplings, further symmetry breaking can take place through conventional Higgs mechanism so that the desired low energy symmetry and the particle spectrum as observed survive in the weak scale.

Here we shall first discuss the symmetry structure as well as the particle identifications just after compactification. We shall then go over to obtain the Yukawa couplings for the cubic superpotentials. It is clear that the post- compactification physics shall contain the effects of the heavy modes in Planck scale that is inherent in models of this type. These modes, when eliminated, will naturally give rise to higher dimension operators in the Lagrangian or the superpotential which will be nonrenormalisable, and shall be suppressed by inverse of Planck scale. These will be written down using only the symmetry structure of the manifold. This symmetry structure will also be exploited for the calculation of the cubic couplings as well, which in fact enables us to write the cubic Yukawa couplings in terms of a single parameter. The forms of the cubic and quartic superpotentials will be exploited to obtain possible low energy scenario consistent with observations. The present section will be a very brief introduction to some of the materials in Ref.[15-20].

12.4.1 Symmetry after Compactification

The Tian Yau manifold is taken as the space $CP^3 \times CP^3$ with x_i and y_i as the homogeneous complex coordinates and the following polynomial conditions:

$$P_1 \equiv \frac{1}{3} \sum_{i=0}^{3} x_i^3 = 0, \qquad (12.50a)$$

$$P_2 \equiv \sum_{i=0}^{3} c_i x_i y_i = 0, \qquad (12.50b)$$

$$P_3 \equiv \frac{1}{3} \sum_{i=0}^{3} y_i^3 = 0. \qquad (12.50c)$$

There is also the discrete symmetry Z_3 which we shall soon give. The equations (12.50) are the same as the equations (A.51) and (A.52) of appendix A, except that the equations in the appendix were more general since we were interested in all the complex deformations and that in equations (12.50) the homogeneous coordinates run from 0 to 3 instead of from 1 to 4. Further, in equation (12.50) we have taken the polynomials so as to anticipate some symmetry properties for the manifold K_0 to explain the methodology and yet obtain sufficiently general results. K_0 defines a six dimensional connected manifold. We then identify different points of K_0 as equivalent with the discrete transformation group Z_3 given by, with $\alpha^3 = 1$,

$$g \equiv diag(1, \alpha^2, \alpha, \alpha) \times diag(1, \alpha, \alpha^2, \alpha) \qquad (12.51)$$

acting on the eight homogeneous coordinates x_i and y_i. Equation (12.51) is the same as equation (A.53) of Appendix A. We have restricted the parameters in equation (12.50) so that the polynomials have Z_3 symmetry. The internal compact space is given as $K = K_0/Z_3$ which is the Tian Yau manifold. Symmetry breaking here will take place through Wilson loops with residual symmetry as a subgroup of E_6 which we now study.

We shall now consider in particular the situation where the residual symmetry group is G_{333} after symmetry breaking takes place through Wilson loops. Using e.g. equation (12.48), and the Dynkin labels for the first nine nonzero roots of 78 of E_6 in Table 1 of Appendix B for the masslessness of the vector mesons for G_{333}, we see that a condition which ensures the residual group to be G_{333} is

12.4. Three Generation Models

$$d = 0 \quad \text{and} \quad a + 2c = 0. \tag{12.52}$$

Parallel to the consideration earlier with Z_3 symmetry group, these equations are true modulo 3, so that e.g. $a + 2c = 0$ is equivalent to $2a + c = 0$. In fact we may identify $exp(2\pi i a/3) = \alpha$, the cube root of unity used in equation (12.51) for the definition of the manifold K. In this case $T(g)$ of equation (12.47) can be identified with

$$T(g) = \begin{pmatrix} 1 & 0 & 0 \\ 0 & 1 & 0 \\ 0 & 0 & 1 \end{pmatrix}_C \times \begin{pmatrix} \alpha & 0 & 0 \\ 0 & \alpha & 0 \\ 0 & 0 & \alpha \end{pmatrix}_L \times \begin{pmatrix} \alpha & 0 & 0 \\ 0 & \alpha & 0 \\ 0 & 0 & \alpha \end{pmatrix}_R. \tag{12.53}$$

The above corresponds to the decomposition

$$27 = (1, \bar{3}, 3) + (3, 3, 1) + (\bar{3}, 1, \bar{3}) \tag{12.54}$$

with a slight change of notation compared to equation (12.44b). Equation (12.53) is the parallel of equation (12.49) considered earlier. There are trivially different choices of a, c and d of equation (12.48) which merely permute the three $SU(3)$ groups above, and we need not consider them separately. We have not yet considered the choice of a, c and d which will give rise to $SU(6) \times U(1)$ as a result of symmetry breaking through Wilson loops parallel to equation (12.52) for G_{333}. We shall do so later.

We shall now recapitulate obtaining the basis of K corresponding to the zero modes considered in subsection 11.3 of Appendix A since besides symmetry, we are interested in the massless sector after compactification. The nine polynomials λ_i ($i = 1, \cdots, 9$) of equation (A.54a) transform under equation (12.51) as independent of a. Hence, corresponding to equation (12.54), we have nine copies of zero modes which are left handed and are color singlets, and thus will consist of lepton multiplets and the Higgs particles. We shall generically call these as "leptons". They are the first nine vectors of Table 3 in Appendix B, as can be explicitly verified by considering their transfromation properties. This can be carried out with equation (12.52) by using the Dynkin labels of Table 3. Each one of them thus has as stated nine copies arising here from the geometrical structure of the internal space.

The zero modes Q_i ($i = 1, \cdots, 7$) in equation (A.54b) each transform like α under equation (12.51). Hence in equation (12.54) they correspond to the symmetry structure (3,3,1). Thus the second set of nine vectors in Table 3 of Appendix B each have seven copies of zero modes arising from the geometry of the internal space. They correspond to the quarks which again can also be verified from the Dynkin labels of the table. Similarly, the zero modes $\bar{Q}_i^c$ ($i = 1, \cdots, 7$) in equation (A.54c) of Appendix A which transform like α^2 under equation (12.51) have the symmetry structure $(\bar{3},1,3)$. Thus the last nine vectors in Table 3 of Appendix B corresponding to the antiquarks each have seven copies of the zero modes coming from internal space. All the above have the same chirality, and are taken as being left handed.

We next consider the basis for $H^{(1,1)}$ which correspond to the mirror particles. The derivations of the same have been omitted in the appendix, and shall be omitted here. We only remark that from 27^* of E_6 there will be six copies $\bar{\lambda}_i$ $(i=1,\cdots,6)$ of the zero modes which which will have opposite chirality and are like mirror leptons, and that there will be four copies each of $\bar{Q}_i$ $(i=1,\cdots,4)$ and $\bar{Q}_i^c$ $(i=1,\cdots,4)$ of mirror quarks and mirror antiquarks. Hence ultimately we can have three generations of light chiral leptons and quarks when the remaining modes become heavy. Just after compactification, we have considered here the residual symmetry group G_{333} which is very large and has too many copies of quarks and leptons arising from the geometry of K. The compactification scale is expected to be very near the Planck scale. We should therefore have further symmetry breaking to generate the Salam Weinberg group as the residual symmetry group, as well as to restrict the number of zero modes to three generations as observed. Six copies of "leptons" and four each of quarks and antiquarks must become massive along with the mirror partners with opposite chirality. We shall now consider this through the construction of Yukawa couplings and conventional Higgs mechanism and see how this can happen.

12.4.2 Yukawa Couplings and Symmetries

The interactions in four dimensions are obtained after one integrates over the compact space K. Yukawa couplings are also naturally obtained by the same procedure. If A, B, C are elements of $H_1(T)$ corresponding to specific superfields, and Ω is the $\bar{\partial}$ closed holomorphic three form, then the corresponding "raw" Yukawa coupling is given by [15]

$$\kappa = \int_K A \wedge B \wedge C \wedge \Omega. \qquad (12.55\text{a})$$

This has next to be corrected for the normalisations of the different fields given through the integrals

$$N(A,B) = \int_K A \wedge *B \qquad (12.55\text{b})$$

where we have the scalar product through the Hodge $*$ operation as in equation (A.37) of Appendix A. The evaluation of the above type

12.4. Three Generation Models

of integrals even for the simplest of relevant Calabi Yau spaces is next to impossible. However, one can obtain relationships between different Yukawa couplings through symmetry relations of K. One also uses the equivalence of different polynomial basis as inherent in the construction of the manifold through polynomial intersections considered by Candelas [18]. We now proceed to state and exploit these properties.

Let us define the following discrete transformations in the manifold K_0 operating on the homogeneous coordinates x and y as below:

$$A = diag(\alpha, \alpha, 1, 1) \times diag(\alpha^2, \alpha^2, 1, 1), \qquad (12.56a)$$

$$B = diag(1, 1, \alpha, 1) \times diag(1, 1, \alpha^2, 1). \qquad (12.56b)$$

Clearly A and B are symmetries of the manifold K_0. Further, they are honest symmetries, which are defined as symmetries of K_0 which descend to K. If all $c_i \neq 0$, then $Z_3(A) \times Z_3(B)$ is a symmetry of the manifold K. We shall consider the transformation properties of the basis vectors under the above transformations. Before doing this, however, let us define some further discrete transformations in K_0:

$$C = diag(1, 1, \sigma) \times diag(1, 1, \sigma), \qquad (12.57a)$$

$$P = diag(\sigma, 1, 1) \times diag(\sigma, 1, 1), \qquad (12.57b)$$

$$S = \text{interchange of } x_i \text{ and } y_i. \qquad (12.57c)$$

In the above, $\sigma = \begin{pmatrix} 0 & 1 \\ 1 & 0 \end{pmatrix}$, such that in the above equations C stands for interchange of last two homogeneous coordinates x_2 and x_3 and y_2 and y_3, and similarly P stands for the interchange of the first two homogeneous coordinates.

In equation (A.54) we have considered a polynomial basis of $H_1(T)$ for $H^{(2,1)}$ corresponding to the zero modes of the particles. This basis has an equivalence relation [17,18]. Let us write the defining polynomials of equations (12.50) as P_α ($\alpha = 1, \cdot, 3$). Then, the polynomials corresponding to cubic terms in x's correspond to $\alpha = 1$, those containing one x and one y correspond to $\alpha = 2$, and those which are cubic in y's correspond to $\alpha = 3$. We then note that the polynomial q_α is equivalent to the following class of polynomials

$$q_\alpha \equiv q_\alpha + X_i \frac{\partial P_\alpha}{\partial x_i} + Y_i \frac{\partial P_\alpha}{\partial y_i} + f_{\alpha\beta} P_\beta. \qquad (12.58)$$

In the above X_i and Y_i are linear functions of the homogeneous coordinates x_i and y_i [18]. Using equations (12.58) we can *write* the polynomials like $x_2 y_3$ as a cubic in x as follows:

$$\begin{bmatrix} 0 \\ x_2 y_3 \\ 0 \end{bmatrix} = \begin{bmatrix} 0 \\ x_2 y_3 \\ 0 \end{bmatrix} + X_i \frac{\partial}{\partial x_i} \begin{bmatrix} P_1 \\ P_2 \\ P_3 \end{bmatrix}. \qquad (12.59)$$

In the above we take $X_3 = -x_2/c_3$ for $c_3 \neq 0$ and take the remaining of the functions as zero. Equation (12.59) with an obvious interpretation then yields that $x_2 y_3 \equiv -x_2 x_3^2 / c_3$.

In order to illustrate the method, let us consider the relatively simple case with

$$c_0 = c_1 = 1 \quad \text{and} \quad c_2 = c_3 = (1 + \epsilon) \qquad (12.60)$$

as has been taken in literature. In that case, in addition to A and B, in equations (12.57) C, P and S also become honest symmetries. With a specific basis we may see that P operates as matter parity, but, it will be clear that other choices for matter parity [20,21] shall also be available. The physics content and also the symmetries will depend on these coefficients, and presumably the preferred manifold can be determined by some process of minimisation in ten dimensions. About this however at present we have very little understanding. We shall therefore mostly use symmetry properties even when we use equations (12.55) for the calculation of Yukawa couplings.

We now use equations (12.58) and take the zero mode "copies" for the lepton, quark and antiquark sectors with the basis of equation (A.54) now replaced by cubic expressions given as

$$\lambda_1 = x_0 x_1 x_2, \ \lambda_2 = x_0 x_1 x_3, \ \lambda_3 = y_0 y_1 y_2, \ \lambda_4 = y_0 y_1 y_3,$$

$$\lambda_5 = x_2^3 + x_3^3, \ \lambda_6 = x_2^3 - x_3^3, \ \lambda_7 = x_1^3 - x_0^3,$$

$$\lambda_8 = x_2 x_3^2, \ \lambda_9 = x_3 x_2^2, \qquad (12.61a)$$

$$Q_1 = x_1 x_2 x_3, \ Q_2 = y_0 y_2 y_3, \ Q_3 = x_0 x_1^2, \ Q_4 = x_1 x_3^2, \ Q_5 = x_1 x_3^2,$$

$$Q_6 = x_2 x_0^2, \ Q_7 = x_3 x_0^2 \qquad (12.61b)$$

$$Q_1^c = x_0 x_2 x_3, \ Q_2^c = y_1 y_2 y_3, \ Q_3^c = x_1 x_0^2, \ Q_4^c = x_0 x_2^2, \ Q_5^c = x_0 x_2^2,$$

$$Q_6^c = x_2 x_1^2, \ Q_7^c = x_3 x_1^2. \qquad (12.61c)$$

We have given the appropriate transformation properties for the above basis corresponding to the discrete symmetries as in Table 1 for

12.4. Three Generation Superstring Models

Table 1

Transformation properties of λ, Q and Q^c and $\bar{\lambda}$, $\bar{Q}$ and $\bar{Q}^c$

Leptons	λ_1	λ_2	λ_3	λ_4	λ_5	λ_6	λ_7	λ_8	λ_9
A:	α^2	α^2	α	α	1	1	1	1	1
B:	α	1	α^2	1	1	1	1	α	α^2
C:	λ_2	λ_1	λ_4	λ_3	1	-1	1	λ_9	λ_8
P:	1	1	1	1	1	1	-1	1	1

Quarks	Q_1	Q_2	Q_3	Q_4	Q_5	Q_6	Q_7
A:	α	α^2	1	α	α	α^2	α^2
B:	α	α^2	1	α^2	1	α	1
C:	Q_1	Q_2	Q_3	Q_4	Q_5	Q_7	Q_6
P:	Q	$\to$	Q^c				

Antiquarks	Q_1^c	Q_2^c	Q_3^c	Q_4^c	Q_5^c	Q_6^c	Q_7^c
A:	α	α^2	1	α	α	α^2	α^2
B:	α	α^2	1	α^2	1	α	1
C:	Q_1^c	Q_2^c	Q_3^c	Q_4^c	Q_5^c	Q_7^c	Q_6^c
P:	Q^c	$\to$	Q				

Mirror Leptons	$\bar{\lambda}_1$	$\bar{\lambda}_2$	$\bar{\lambda}_3$	$\bar{\lambda}_4$	$\bar{\lambda}_5$	$\bar{\lambda}_6$
A:	1	1	α	α	α^2	α^2
B:	1	1	α	α	α^2	α^2
C:	1	1	-1	-1	-1	-1
P:	1	1	-1	-1	-1	-1

MirrorQuarks	$\bar{Q}_1$	$\bar{Q}_2$	$\bar{Q}_3$	$\bar{Q}_4$	$\bar{Q}_1^c$	$\bar{Q}_2^c$	$\bar{Q}_3^c$	$\bar{Q}_4^c$
A:	1	1	1	1	1	1	1	1
B:	α	α^2	α	α^2	α	α^2	α	α^2
C:	$\bar{Q}_2$	$\bar{Q}_1$	$\bar{Q}_4$	$\bar{Q}_3$	$\bar{Q}_2^c$	$\bar{Q}_1^c$	$\bar{Q}_4^c$	$\bar{Q}_3^c$
P:	$\bar{Q}$	$\leftrightarrow$	$\bar{Q}^c$					

leptons λ, quarks Q and antiquarks Q^c of 27 and also for the mirror set $\bar{\lambda}$, $\bar{Q}$ and $\bar{Q}^c$ of 27^*. These, or the definitions of the transformations giving the honest symmetries, are then utilised for the evaluation of the Yukawa couplings which in turn gives the phenomenology at intermediate scales or at low energy.

The nonvanishing "raw" Yukawa couplings defined through equation (12.55a), which are invariant under honest symmetries, are given in terms of the following polynomials of degree nine:

$$B_1 \equiv x_i^9;\ B_2 \equiv x_m^9;\ B_3 \equiv x_i^6 x_j^3;\ B_4 \equiv x_i^6 x_m^3;\ B_5 \equiv x_m^6 x_n^3;$$

$$B_6 \equiv x_m^6 x_i^3;\ B_7 \equiv x_m^3 x_n^3 x_i^3;\ B_8 \equiv x_i^3 x_j^3 x_m^3;$$

$$B_9 \equiv x_i^3[x_i x_j x_m][y_i y_j y_m];\ B_{10} \equiv x_m^3[x_i x_j x_m][y_i y_j y_m];$$

$$B_{11} \equiv x_n^3[x_i x_j x_m][y_i y_j y_m];\ B_{12} \equiv x_i^3[x_i x_m x_n][y_i y_m y_n];$$

$$B_{13} \equiv x_j^3[x_i x_m x_n][y_i y_m y_n] B_{14} \equiv x_n^3[x_i x_m x_n][y_i y_m y_n]. \quad (12.62)$$

In the above, we have taken $i, j = 0, 1$ and $m, n = 2, 3$, and that the B_a's ($a = 1, \cdots, 14$) are the Yukawa couplings of the superfields. The corresponding integrals as in equation (12.55a) are impossible to evaluate. But as already stated we note that the integrals are related to each other through equations like (12.58). The relationships between these Yukawa couplings for some manifolds have been derived and utilised in Ref.[16-19]. The nonvanishing cubic Yukawa couplings in the λ sector are given as [17-19]

$$\lambda_1^3,\ \lambda_2^3,\ \lambda_3^3,\ \lambda_4^3,\ \lambda_5^3,\ \lambda_6^3,\ \lambda_7^3,\ \lambda_8^3,\ \lambda_9^3,$$

$$\lambda_1 \lambda_3 \lambda_5,\ \lambda_1 \lambda_3 \lambda_6,\ \lambda_2 \lambda_4 \lambda_5,\ \lambda_2 \lambda_4 \lambda_6,\ \lambda_1 \lambda_4 \lambda_9,\ \lambda_2 \lambda_3 \lambda_8,$$

$$\lambda_6^2 \lambda_5,\ \lambda_6^2 \lambda_7,\ \lambda_8 \lambda_9 \lambda_5. \quad (12.63)$$

With $\epsilon = 0$ in equation (12.60), we also quote here the nonvanishing cubic couplings containing the color singlet λ's and quark and antiquark. These are,

$$\lambda_2 Q_7 Q_7^c \equiv 3,\ \lambda_2(Q_2 Q_6^c + Q_6 Q_2^c) \equiv 1,\ \lambda_2(Q_5 Q_3^c + Q_3 Q_5^c) \equiv 1,$$

12.4. Three Generation Models

$$\lambda_3(Q_1Q_5^c + Q_5Q_1^c) \equiv 1, \quad \lambda_3(Q_3Q_6^c + Q_6Q_3^c) \equiv 1, \quad \lambda_3Q_4Q_4^c \equiv 3,$$

$$\lambda_4(Q_3Q_7^c + Q_7Q_3^c) \equiv 1, \quad \lambda_4(Q_1Q_4^c + Q_4Q_1^c) \equiv 1, \quad \lambda_4Q_5Q_5^c \equiv 3,$$

$$\lambda_6(Q_4Q_6^c + Q_6Q_4^c) \equiv -2, \quad \lambda_6(Q_5Q_7^c + Q_7Q_5^c) \equiv -2,$$

$$\lambda_8(Q_4Q_7^c + Q_7Q_4^c) \equiv 1, \quad \lambda_8(Q_2Q_5^c + Q_5Q_2^c) \equiv 1,$$

$$\lambda_8(Q_1Q_6^c + Q_6Q_1^c) \equiv 1, \quad \lambda_9(Q_5Q_6^c + Q_6Q_5^c) \equiv 1,$$

$$\lambda_9(Q_1Q_7^c + Q_7Q_1^c) \equiv 1, \quad \lambda_9(Q_2Q_4^c + Q_4Q_2^c) \equiv 1. \quad (12.64)$$

We note that the above Yukawa couplings will give mass to the extra colored "quarks" which mediate fast proton decay, when some of the color singlet λ's attain vacuum expectation values. They are thus phenomenologically more relevant.

The quartic Yukawa couplings in superspace involve nonrenormalisable terms in the Lagrangian. They arise from the elimination of the heavy modes of the ten dimensional superstring theory. There are a large number of such modes, and hence effective quartic couplings shall be an inevitable part of the phenomenology. The only constraint we can utilise for them with unknown dynamics of compactification is to consider those quartic couplings which remain invariant under honest symmetries. The quartic couplings are taken with ordinary supermultiplets as well as the mirror super- multiplets with the symmetry transformations as in Table 1. We shall consider them occasionally as and when they become relevant for the consideration of symmetry breaking.

12.4.3 Intermediate Scales

We have noted that after compactification, the symmetry group is G_{333}. The multiplet structure is given by equation (12.54) and Table 3 of appendix B. We now give the group structure for the same in the context of symmetry breaking at intermediate scales. For the λ's of $(1, \bar{3}, 3)$ of $SU(3)_C \times SU(3)_L \times SU(3)_R$ we have,

$$\lambda = \begin{bmatrix} H^0 & H^+ & e^c \\ \tilde{H}^- & \tilde{H}^0 & \nu^c \\ e^- & \nu & N \end{bmatrix} = \begin{bmatrix} H_u & e^c \\ H_d & \nu^c \\ L & N \end{bmatrix}. \quad (12.65a)$$

Similarly, for Q of $(3,3,1)$ and Q^c of $(\bar{3},1,\bar{3})$ we have as in Table 3 of Appendix B,

$$Q = \begin{bmatrix} u \\ d \\ D \end{bmatrix} = \begin{bmatrix} q \\ D \end{bmatrix}; \quad Q^c = \begin{bmatrix} u^c \\ d^c \\ \tilde{D} \end{bmatrix} = \begin{bmatrix} q^c \\ \tilde{D} \end{bmatrix} \qquad (12.65b)$$

We note that in the matrix in equation (12.65a), $\bar{3}$ of $SU(3)_L$ is horizontal, and conventional $SU(2)_L$ is read from right to left. In (12.65b), 3 of $SU(3)_L$ is vertical, and the color indices are suppressed. Also,

$$Q = I_{3L} + I_{3R} + \frac{1}{2}(B-L). \qquad (12.66)$$

After compactification we have as noted earlier 9+7+7 copies of leptons, quarks, and antiquarks, and, 6+4+4 copies of the mirror fields as in Table 1. While symmetry breaking takes place, most of these must become heavy for the theory to agree with observations. Further, the following are also to be ensured. Firstly, rapid proton decay must be avoided. Since D and $\tilde{D}$ mediate proton decay, while coming down to three generations, these must become much heavier than the TeV scale. Light Higgs doublets must also be available for weak symmetry breaking. Further, $sin^2\theta_W$ must have the experimental value at low energies consistent with renormalisation group equations. Flavor changing neutral currents should also be absent. Without giving details, we shall generally see how such things can happen.

We shall now examine in an ad hoc fashion the consequences when some of the fields have nonzero vacuum expectation values. For subsequent symmetry breaking, let us take the following fields to have nonzero vacuum expectation values at intermediate scales V_0 and V_{B-L} given as (Kalara and Mohapatra, [17]):

$$(\lambda_9, \lambda_2, \lambda_3, \bar{\lambda}_3, \bar{\lambda}_4)_{(1,1)} \equiv V_0, \qquad (12.67a)$$

$$(\lambda_8, \lambda_4, \lambda_6, \bar{\lambda}_1, \bar{\lambda}_2)_{(1,16)} \equiv V_{B-L}. \qquad (12.67b)$$

In the above equations, the subscripts as before refer to the copies due to compactification as in equation (12.63) and the subscripts within parenthesis give the $SU(5)$ and $SO(10)$ representations respectively corresponding to the fields N and ν^c as in Table 3 of Appendix B.

Let us first consider the masses in the leptonic sector. Firstly, λ_7 has odd P parity, and thus remains massless. Thus, $H_{u,7}$, $H_{d,7}$, L_7 and e_7^c are massless. Next we attempt to construct the mass matrix when there is mixing. Taking into account $SU(3)_L$ and $SU(3)_R$ invariance, for the $SU(2)_L$ doublets we can write the mass matrix as

$$L_{\text{mass}} = \begin{bmatrix} H_{u,5} & H_{u,6} \end{bmatrix} M \begin{bmatrix} H_{d,5} \\ L_5 \\ H_{d,6} \\ L_6 \end{bmatrix}. \qquad (12.68)$$

12.4. Three Generation Models

In the above, M is a 4×2 matrix with the matrix elements obtained from equation (12.67) multiplied by the respective Yukawa couplings as in (12.63). The above mass contribution in general leads to two lepton doublets which remain massless. We shall write them as

$$\psi_2 = p_1 L_5 + p_2 L_6 + p_3 H_{d,5} + p_4 H_{d,6}, \qquad (12.69a)$$

$$\psi_3 = p'_1 L_5 + p'_2 L_6 + p'_3 H_{d,5} + p'_4 H_{d,6}. \qquad (12.69b)$$

The coefficients p_i and p'_i $(i = 1, \cdots, 4)$ can be calculated when M is explicitly known. Thus at low energies we have three lepton doublets given as $\psi_1 = L_7$ and ψ_2 and ψ_3. Also, the symmetry properties and the Yukawa couplings further give us that the right handed charged leptons e^c_5, e^c_6 and e^c_7, and the two Higgs doublets $H_{u,7}$ and $H_{d,7}$ remain massless as needed.

Let us now see the possible mode of generation of the intermediate scales. We give here a simplified version of the mechanism. Let us consider a single complex scalar field ϕ with a soft supersymmetry breaking term in the Lagrangian, along with the dimension six operators from the quartic superpotential quoted earlier, and with the fields eliminated having masses in the compactification or Planck scale. We then have

$$L \approx -M_s^2 \phi^* \phi + (f/M_c)^2 (\phi^* \phi)^3, \qquad (12.70)$$

where f is a dimensionless parameter. In the above, M_s is the supersymmetry breaking scale of the order of a TeV, and M_c is the compactification scale $\geq 10^{18}$ GeV. An exrtremisation then yields that

$$<\phi> \approx f^{-1/2}(M_s M_c)^{1/2} = M_I \geq 10^{11} \ GeV. \qquad (12.71)$$

The above is the intermediate scale we are talking about. The actual value depends on f and thus will depend on the details of dynamics and could as well be much smaller than unity.

With the light leptonic sector, again invoking the quartic couplings, we can show that through seesaw mechanism ν^c_7, ν^c_5, N_1, N_2 attain masses of the order of M_I^2/M_c, so that they are not observable at present energies. This detail is model dependant and is being omitted here.

We now consider the question of D and $\tilde{D}$ quarks becoming heavy. In fact, the mass term of the superpotential obtained from equation (12.64) becomes

$$L_{mass} = 3V_0 \Big[D_2 \tilde{D}_2 + D_2 \tilde{D}_6 + D_6 \tilde{D}_2 + D_3 \tilde{D}_5 + D_5 \tilde{D}_3$$

$$+ \ D_1\tilde{D}_5 + D_5\tilde{D}_1 + D_1\tilde{D}_4 + D_4\tilde{D}_1 - \frac{1}{3}\Big\{D_5\tilde{D}_6 + D_6\tilde{D}_5$$
$$+ \ D_7\tilde{D}_1 + D_1\tilde{D}_7 + D_2\tilde{D}_4 + D_4\tilde{D}_2\Big\}\Big]$$
$$- \ V_{B-L}\Big[D_4d_1 + D_1d_4 + D_3d_7 + D_7d_3 + D_4d_7$$
$$+ \ D_7d_4 + D_5d_7 + D_7d_5 + D_2d_5 + D_5d_2\Big]. \qquad (12.72)$$

Thus, proton decay calculations [24] need not give rise to an unacceptable value, since all the D and $\tilde{D}$ masses can become high. In fact, proton decay takes place through the dimension six operators corresponding to the process of two quarks going to antiquark plus (anti)lepton. For a massive D or $\tilde{D}$ quark as given by equation (12.72), the corresponding diagrams will partly consist of operators with a lepton, an antiquark, and two scalar quarks as external legs, and thus will be divided by the mass m_D of D since this contribution arises from the heavy D or $\tilde{D}$ "quark" propagator. This diagram is now to be dressed with the exchange of a Wino, Zino and gluino and the propagators for the two scalar quarks in a triangle diagram to yield the effective contribution for proton decay referred to above. Hence these dimension six operator for proton decay will be suppressed by the product of m_D and a mass around the supersymmetry breaking scale. The operator thus for the high scales is effectively a dimension five operator. This has been considered e.g. in Ref [25] where a lower limit of the mass m_D is derived as a function of the scalar quark mass as well as the photino mass. Clearly, if the scalar quark mass is sufficiently large, say around 500 GeV, m_D could be reasonably "small", say around 10^{14} GeV [25]. It has been further pointed out [26] that the symmetry structure may also substantially inhibit proton decay when the mixings are small, and therefore the observed stability of the proton could be understood more easily.

We have taken the above example to illustrate the phenomenological implications for compactification of higher dimensional supergravity through complex manifolds. For what we have considered, it can be verified that the supersymmetry breaking scale does not become satisfactory. Neutrino mass however becomes "satisfactory" in the sense that at the tree level we have one massless neutrino and through seesaw mechanism, two ultralight neutrninos [17].

Another type of attempt for phenomenology has been that of Ref. [20], where one defines matter parity through the transformation C of equation (12.57a) instead of taking it as P of (12.57b). As an element

12.4. Three Generation Models

of G_{333} this becomes [21]

$$U_z = diag[1,1,1] \times diag[-1,-1,-1] \times diag[-1,-1,-1]. \quad (12.73)$$

For the basis in compact space as in equation (12.61), the corresponding transformation properties are already quoted in Table 1. This definition of matter parity plays a crucial role in the analysis of Ref. [21], and one sees that a high intermediate mass scale $\approx 10^{15}$ GeV can be generated for the vacuum expectation values of both ν^c and N. We note that M_c as nearly being the Planck scale is here assumed, which is expected since the dynamics of compactification is likely to be closely linked with the string dynamics.

In the examples cited, one takes matter parity as given by a Z_2 group. However, for the definition of matter parity a group Z_3 can also be chosen, which changes the phenomenological picture. Here in fact one uses the B symmetry of equation (12.56b) to define matter parity [19,21]. The corresponding phenomenology has been discussed in some detail in [19], which appears as reasonable, except for the neutrino masses, a situation which may however be circumvented if we have E_6 singlets.

We note that in all the attempts so far with G_{333} as the symmetry group after compactification the phenomenology is not very satisfactory either at the level of neutrino masses or of breaking of supersymmetry or proton decay while ensuring that only the observed particles are there below the weak scale. The investigations are thus incomplete and there are many places where we feel that we do not know enough. Clearly more effort is called for to link Planck scale with the weak scale in a quantitative manner. The problem however seems as tractable and need not be left for the "future".

Besides G_{333} being the residual symmetry group after symmetry breaking through Wilson loops, we can have also other group structures. We briefly note some of these in the next subsection.

12.4.4 Compactification with E_6 going to $SU(6) \times U(1)$

We recall that the equation (12.52) corresponds to the symmetry group after compactification being given as G_{333}. The symmetry breaking direction consistent with G_{321} at low energy gave rise to equation (12.48) for the Dynkin labels in dual basis. Equation (12.52) ensured that all the masses of the gauge bosons of G_{333} are zero through equation (B.16). In case a, c and d in equation (12.51) are such that all the gauge bosons of the group $SU(6) \times U(1)$ are zero from the Dynkin labels in Table 1 of Appendix B, then the residual gauge symmetry after compactification shall be $SU(6) \times U(1)$ or bigger. We note that flux breaking of symmetry shall not change the rank of the group, so that here we shall have a group of rank six. We give below eleven equations for a, c, and d in (12.48) so that the residual group in fact becomes the $SU(6) \times U(1)$ group. These are [16],

$$d = 0, a = 0; \quad c = d, a = 0; \quad d = 2c, a + 2c = 0;$$

$$c = d, a + 2c = 0; \quad c = 0, a = 0; \quad c = d, a = 0;$$

$$c = 0, a + d = 0; \quad c = d, a + 2d = 0; \quad c = 0, d = 0;$$

$$a + c = 0, d = 0; \quad a + c = 0, 2a + d = 0. \tag{12.74}$$

As before, all the equalities here are modulo 3 corresponding to the discrete group Z_3. We may also remark that if we take $c = 0$ and $a + 2d = 0$, then E_6 symmetry will remain unbroken [16]. These will have the standard identification of weak isospin and charge as in equation (B.23a), or "flipped" identification as in equation (B.23b). The phenomenology of this with a flipped representation of the particle spectra has been discussed in [22,23]. These attempts in the context of compactification and the calculation of the Yukawa couplings are however at a preliminary stage.

12.5 Discussions

In this Chapter we have considered various aspects of supersymmetry theory and phenomenology based on higher dimensional theories, including in particular what might be expected from superstring theories. The mathematical as well as the physical concepts involved are beautiful, but success regarding plhenomenology is limited. At present we shall consider some aspects where further generalisation of the ideas may be possible, or, insight gained.

We note that the conventional picture for spontaneous symmetry breaking is through Higgs mechanism where an elementary scalar field attains a vacuum expectation value (vev). On the other hand, we expect that gluino condensate mechanism for E_8' in $E_8 \times E_8'$ based on heterotic string as discussed in section 11.3 indicates that there the condensate mechanism shall breaks E_8' symmetry. In this Chapter in fact we have only considered E_8 symmetry breaking. Can we use the condensate mechanism which is familiar to us but is more troublesome

12.5. Discussions

to tackle? In section 10.4 e.g. we noticed that it might permit the Higgs doublets to be superheavy. In fact when we recall equation (10.44b) as

$$<\phi> = <J_\phi>/m_\phi^2, \qquad (12.75)$$

with $<J_\phi> = \mu^3$ as the vev for the condensate, the Higgs particle mass can be much larger than its vacuum expectation value, and mass and vev do not get linked up as in $\lambda\phi^4$ type of symmetry breaking. Based on superpotentials as discussed in section 12.2.2, the problem of accommodating superheavy Higgs particles in supergravity through condensates has been discussed in Ref. [27]. One takes the condensates of fermions which are already heavy and have a Dirac mass, so that with this mechanism no chiral symmetry breaking for these fermions is involved. On the other hand in technicolor type of theories one takes condensates of light fermions parallel to conventional chiral symmetry breaking, avoided in the above mechanism [27]. Similarly condensates of heavy fermions were invoked for the consideration of light Dirac neutrinos in a superstring motivated supergravity model with E_6 symmetry [28]. Can we invoke the mechanism through condensates to replace equations (12.67) for generating the intermediate symmetries or symmetry breaking? For this purpose, the approaches of [27,28] must be supplemented by the sophisticated technology of the present Chapter to include the structure of the compact space, and it may be worthwhile to do so. Then, some of the constraints for model construction will become less severe.

In the present Chapter we have based our discussions only on the low energy supergravity limit of superstring theory. There are many heavy modes in superstring theories which should percolate into supergravity as a "low energy" theory and these should generally speaking vitiate models for compactification which is expected to take place around the Planck scale. Therefore compactifications based on string theory may contain more reliable information than what is done here. It is reassuring to note that generally speaking string theory compactifications give very similar results to what has been considered here. E.g. with (p,q) supersymmetry in two dimensions, about which we do not discuss in the pressent text, one can discuss compactification with similar outputs [29]. Gepner also shows that string compactification corresponds to the same on exactly solvable Calabi Yau spaces [30]. This shows that string dynamics probably chooses the structure of space, and the simple way in which we have considered the same need have only minor modifications even when the details of dynamics in string space are included. String theory is nice because here most of the dynamics gets converted to symmetries in two dimensions such as modular invariance, which simplifies the problem.

Although less attractive in the context of the prejudice for string theory, compactification can also be considered with coset space formulation for the geometry of the compact space, and phenomenology in supergravity has also been examined around this [31].

Some further work has been done by Scrimmrigk regarding the construction of three generation Calabi Yau manifolds [32], which could be relevant for phenomenology [26]. Further work in this context to clarify whether it gives better results than earlier model constructions is needed.

We end this Chapter with a brief comment regarding higher dimensional spaces and compactification in the context of symmetry. Initially Kaluza Klein theories were constructed with the objective of generating symmetry from the geometry of the extra dimensios. But in the present Chapter our obsession has been to lose symmetry during compactification, which is justified under the circumstances. We would like to call such theories with symmetry breaking arising from the extra dimensions as "anti-Kaluza Klein" theories. Is it useful to have a higher dimensional theory where symmetry is partly gained, and partly lost, and will any such theory be relevant?

REFERENCES

1. M.B. Green and J.H. Schwarz, Phys. Lett. 149B, 117 (1984).

2. A.H. Chamseddine, Nucl. Phys. B185, 403 (1981).

3. C.S. Chapline and N.S. Manton, Phys. Lett. 120B, 105 (1983).

4. E. Witten, Phys. Lett. 155B, 151 (1985).

5. P. Candelas, G.T. Horowitz, A. Strominger and E. Witten, Nucl. Phys. B258, 46 (1985).

6. P. Candelas and D.J. Reine, Nucl. Phys. B248, 415 (1984)

7. Y. Hosotani, Phys. Lett. 129B, 193 (1984).

8. B.A. Ovrut, Prog. Theor. Phys. 86, 185 (1986); K. Choi and J.E. Kim, Phys. Lett. 176B, 103 (1986).

9. A.M. Polyakov, Phys. Lett. 59B, 82 (1975).

10. R. Slansky, Phys. Rep. 79, 1 (1981).

11. E. Witten, Nucl. Phys. B258, 75 (1985); M. Dine, V. Kaplunovsky, M. Mangano, C. Nappi and N. Seiberg, Nucl. Phys. B258, 519 (1985).

12.5. Discussions

12. J. Breit, B. Ovrut and G. Segre, Phys. Lett. 158B, 33 (1985); F. Del Aguila, G. Blair, M. Daniel and G.G Ross, Nucl. Phys. B272, 413 (1986).

13. J.P. Derendiger, L. Ibanez and H.P. Nilles, Nucl. Phys. B267, 365 (1986).

14. E. Gildner, Phys. Rev. D14, 1667 (1976); S.M. Barr, Phys. Lett. 112B, 219 (1982).

15. A. Strominger, Phys. Rev. Lett. 55, 2547 (1985); A. Strominger and E. Witten, Commun. Math. Phys. 101, 341 (1985).

16. B. Greene, K. Kirklin, P. Miron and G.G. Ross, Nucl. Phys. B278, 667 (1986); B292, 606 (1987).

17. S. Kalara and R.N. Mohapatra, Phys. Rev. D36, 3474 (1987); S. Kalara, P.K. Mohapatra and R.N. Mohapatra, Phys. Rev. D 37, 3284 (1988); R.N. Mohapatra, E. Rusjan, G. Senjanovic and A. Sokorac, Phys. Lett. 225B, 73 (1989).

18. P. Candelas and S. Kalara, Nucl. Phys. B298, 357 (1988); P. Candelas, Nucl. Phys. B298, 458 (1988); P. Candelas, A.M. Dale, C.A. Lutkon and R. Schimmrigk, Nucl. Phys. B298, 493 (1988).

19. G. Lazarides and Q. Shafi, Nucl. Phys. B308, 451 (1988); G. Lazarides, P.K. Mohapatra, C. Panagiotakopoulos and Q. Shafi, Nucl. Phys. B323, 614 (1989).

20. P. Nath and R. Arnowitt, Phys. Rev. Lett. 62, 1437 (1989); Phys. Rev. 39, 2006 (1989).

21. M. Bento, H. Hall and G.G. Ross, Nucl. Phys. B292, 400 (1987).

22. B. Campbell, J. Ellis, J.S. Hagelin, D.V. Nanopoulos and R. Ticciati, Phys. Lett. 198B, 200 (1987); I. Antoniadis, J. Ellis, J.S. Hagelin and D.V. Nanopoulos, Phys. Lett. 208B, 209 (1988).

23. C. Panagiotakopoulos, Bartol preprint BA-89-20 (1989); CERN-TH 5516/89 (1989).

24. K. Biswal, L. Maharana and S.P. Misra, Phys. Rev. D25, 266 (1982).

25. R. Arnowitt and P. Nath, Phys. Rev. Lett. 60, 1817 (1988)

26. C. Panagiotakopoulos, CERN-TH.5516/89 (1989).

27. S. Mahapatra and S.P. Misra, Phys. Rev. D33, 3464 (1986).

28. S. Mishra and S.P. Misra, Phys. Lett. 217B, 65 (1989).

29. C.M. Hull and E. Witten, Phys. Lett. 160B, 398 (1985); J.J. Atick, L.J. Dixon and A. Sen, Nucl. Phys. B292, 109 (1987).

30. D. Gepner, Nucl. Phys. B296, 757 (1988); ibid. B311, 191 (1988); J. Distler and B. Gepner, Nucl. Phys. B309, 295 (1988).

31. T.R. Govindarajan, A.S. Joshipura, S.D. Rindhani and U. Sarkar, Phys. Rev. Lett. 57, 2409 (1986).

32. R. Schimmrigk. Phys. Lett. 193B, 175 (1987).

Chapter 13

Supersymmetry in Two Dimensions

13.1 Introduction

Field theory in two dimensions is extremely helpful in recognising the general structure of a theory, as well as getting an intuitive picture regarding different conceptual inputs or outputs. The theory here is not as complicated as in four dimensions, and is not as trivial as in quantum mechanics, which is field theory in $1+0$ dimensions. In our discussions earlier, we had mostly considered dynamics at the tree level, and at zero temperature. On the other hand, it will be useful to recognise the effect of nonperturbative dynamics in field theory as well as that at finite temperature in the context of supersymmetry. With this in view, we shall consider in this Chapter supersymmetry in 1+1 dimensions with a superspace formulation. We shall also illustrate phase transitions as well as the temperature dependance of the same with a nonperturbative variational method as was briefly introduced in section 6 of Chapter 6. The 1+1 dimensional space becomes a very good laboratory to study such theoretical ideas as were omitted earlier.

13.1.1 Spinors in 1+1 dimensions

Here we shall take $\gamma^0 = \sigma_2$ and $\gamma^1 = i\sigma_1$ with $g^{mn} = diag(1,-1)$. The Dirac spinor ψ has two components, each one being complex. We can then define the charge conjugation operation through the matrix C such that

$$\psi^{(c)} = C\bar{\psi}^{tr}. \qquad (13.1)$$

In the above, $\bar{\psi} = \psi^\dagger \gamma^0$. Clearly we may take $C = \sigma_2$. Then, self charge conjugate spinors, or, Majorana spinors, given through the condition $\psi^{(c)} = \psi$ are such that

$$\psi = \psi^*, \qquad (13.2)$$

i.e. in this basis, both the components must be real. Thus, for two Majorana spinors ψ and ξ, the components of which necessarily anticommute, the following equations are true:

$$\bar{\psi}\xi = \bar{\xi}\psi = (\bar{\psi}\xi)^\dagger, \qquad (13.3a)$$

$$\bar{\psi}\gamma^m\xi = -\bar{\xi}\gamma^m\psi, \qquad (13.3b)$$

$$\bar{\psi}\gamma^m\gamma^n\xi = \bar{\xi}\gamma^n\gamma^m\psi, \qquad (13.3c)$$

$$(\bar{\psi}\gamma^m\xi)^\dagger = \bar{\xi}\gamma^m\psi, \qquad (13.3d)$$

and

$$\psi_\alpha \psi_\beta = -\frac{1}{2}(\gamma^0)_{\alpha\beta}\bar{\psi}\psi. \qquad (13.3e)$$

13.1.2 Superspace for 1+1 dimensions

As earlier, we shall proceed here to define supersymmetry through superspace, which we take as (x^m, θ) with θ being a Majorana spinor the components of which are Grassmannian variables. We now define, parallel to equations (2.2) and (4.1) in Chapters 2 and 4 respectively, the supersymmetry transformation corresponding to the present superspace as

$$x^m \to x^m + i\bar{\xi}\gamma^m\theta, \qquad (13.4a)$$

$$\theta_\alpha \to \theta_\alpha + \xi_\alpha. \qquad (13.4b)$$

In the above ξ_α are the parameters for the supersymmetry transformation which are components of a Majorana spinor and are Grassmannian variables. Parallel to equations (2.3) and (4.3), we then clearly have the corresponding generators for the above transformations as differentiation operators given as

$$Q_\alpha = (\gamma^0)_{\alpha\beta}\frac{\partial}{\partial \theta_\beta} + i(\gamma^m)_{\alpha\beta}\,\theta_\beta \frac{\partial}{\partial x^m}, \qquad (13.5)$$

so that the supersymmetric transformations are generated by $\delta_\xi \equiv \bar{\xi}Q$. We have clearly the antisymmetry algebra for Grassmannian variables as a part of the graded Lie algebra (when we include "Lorentz" group) now given as

13.2. Supersymmetry in 1+1 dimensions

$$[Q_\alpha, \bar{Q}_\beta]_+ = 2i(\gamma^m)_{\alpha\beta} \frac{\partial}{\partial x^m}. \qquad (13.6)$$

We can as before define the covariant derivatives which must anticommute with the above generators, so that they in fact commute with the supersymmetry transformations. These become

$$D_\alpha = (\gamma^0)_{\alpha\beta} \frac{\partial}{\partial \theta_\beta} - i(\gamma^m)_{\alpha\beta}\, \theta_\beta \frac{\partial}{\partial x^m}. \qquad (13.7)$$

13.2 Supersymmetry in 1+1 dimensions

We now proceed to formulate supersymmetric field theory in two dimensions. As earlier, we shall do so with the definition of superfields defined on superspace.

13.2.1 Superfields in 1+1 dimensions

We now define the superfield Φ as

$$\Phi(x,\theta) = \phi(x) + \bar{\theta}\psi(x) + \frac{1}{2}\bar{\theta}\theta d(x). \qquad (13.8)$$

In the above, the Φ is a Lorentz scalar, $\phi(x)$ is a real scalar field, $\psi(x)$ is a Majorana spinor, and, as we shall see later, $d(x)$ is an auxiliary field. As earlier, (ϕ, ψ, d) are the component fields of the superfield Φ. To obtain the transformation properties of these component fields under a supersymmetry transformation, we note that,

$$\delta_\xi \Phi = \delta_\xi \phi(x) + \bar{\theta}\delta_\xi \psi(x) + \frac{1}{2}\bar{\theta}\theta \delta_\xi d(x). \qquad (13.9)$$

Using equation (13.5), this yields the transformation properties for the component fields as,

$$\delta_\xi \phi = \bar{\xi}\psi, \qquad (13.10a)$$

$$\delta_\xi \psi = \xi d - i\gamma^m \xi \partial_m \phi, \qquad (13.10b)$$

and,

$$\delta_\xi d = -i\partial_m[\bar{\xi}\gamma^m \psi]. \qquad (13.10c)$$

We note that as earlier, products of superfields will be superfields, and that the $\bar{\theta}\theta$ component of the superfield transforms like a total divergence. These will be utilised to construct the supersymmetric Lagrangian.

13.2.2 Lagrangian Construction

For the construction of the supersymmetric Lagrangian we use the superfield $\Phi(x,\theta)$ along with the covariant derivative (13.7) and write Lagrangian density in superspace for free fields as $L_0 = \frac{1}{2}(\bar{D}_\alpha \Phi)(D_\alpha \Phi)$. Here we also define the "measure" $d^2\theta$ such that

$$\int \bar{\theta}\theta d^2\theta = 1. \tag{13.11}$$

We now note that

$$D_\alpha \Phi = \psi_\alpha + d\theta_\alpha - i\gamma^m{}_{\alpha\beta}\theta_\beta \partial_m \phi + \frac{i}{2}(\bar{\theta}\theta)\gamma^m{}_{\alpha\beta}\partial_m \psi_\beta, \tag{13.12}$$

such that

$$\begin{aligned}\mathcal{L}_0 &= \frac{1}{2}(\bar{D}_\alpha \Phi)(D_\alpha \Phi) \\ &= \frac{1}{2}\bar{\psi}\psi + \bar{\theta}(i\gamma^m \psi \partial_m \phi + d\psi) \\ &\quad + \frac{1}{2}\bar{\theta}\theta \left[i\bar{\psi}\gamma^m \partial_m \psi + (\partial_m \phi)(\partial^m \phi) + d^2\right].\end{aligned} \tag{13.13}$$
$$\tag{13.14}$$

We shall now include interactions by taking the total Lagrangian density in superspace as

$$\mathcal{L} = \mathcal{L}_0 + W(\Phi), \tag{13.15}$$

where we have the expansion that,

$$\begin{aligned}W(\Phi) &= W(\phi) + (\bar{\theta}\psi + \frac{1}{2}d\,\bar{\theta}\theta)\frac{dW(\phi)}{d\phi} \\ &\quad + \frac{1}{2}(\bar{\theta}\psi + \frac{1}{2}d\,\bar{\theta}\theta)^2 \frac{d^2W(\phi)}{d\phi^2}.\end{aligned} \tag{13.16}$$

From equations (13.14) and (13.16), we then substitute that with obvious notations the superfield $\mathcal{L}$ is given as

$$\mathcal{L} = \phi^{(\mathcal{L})} + \bar{\theta}\psi^{(\mathcal{L})} + \bar{\theta}\theta \mathcal{L}^{(\mathcal{L})}. \tag{13.17}$$

Thus the Lagrangian in two dimensions is given as

$$\begin{aligned}L &= \int d^2\theta \mathcal{L} = \mathcal{L}^{(\mathcal{L})} \\ &= L_0 + L_{\text{int}}.\end{aligned} \tag{13.18a}$$

Clearly in the above the free field Lagrangian density L_0 in two dimensions in terms of the component fields is given by

13.2. Supersymmetry in 1+1 dimensions

$$L_0 = \frac{1}{2}\left[i\bar{\psi}\gamma^m\partial_m\psi + \partial_m\phi\partial^m\phi + d^2\right], \quad (13.18b)$$

with d as the auxiliary field, and the interactions part L_{int} is given by

$$L_{\text{int}} = \frac{d}{2}\frac{dW}{d\phi} - \frac{1}{4}\bar{\psi}\psi\frac{d^2W}{d\phi^2}. \quad (13.18c)$$

From the above Lagrangian we obtain the field equations as,

$$d = -\frac{1}{2}\frac{dW}{d\phi}, \quad (13.19a)$$

$$i\gamma^m\partial_m\psi = \frac{1}{2}\psi\frac{d^2W}{d\phi^2}, \quad (13.19b)$$

and,

$$\partial_m\partial^m\phi = \frac{d}{2}\frac{d^2W}{d\phi^2} - \frac{1}{4}\bar{\psi}\psi\frac{d^3W}{d\phi^3}. \quad (13.19c)$$

We note that with the elimination of the auxiliary field the Lagrangian density in two dimensions becomes

$$\begin{aligned}L &= \frac{1}{2}\left[i\bar{\psi}\gamma^m\partial_m\psi + \partial_m\phi\partial^m\phi\right] \\ &- \frac{1}{8}\left(\frac{dW}{d\phi}\right)^2 - \frac{1}{4}\bar{\psi}\psi\frac{d^2W}{d\phi^2}.\end{aligned} \quad (13.20)$$

Also, the Hamiltonian density in two dimensions is given as

$$\begin{aligned}H &= \frac{1}{2}\left[-i\bar{\psi}\gamma^1\partial_1\psi + (\partial_0\phi)^2 + (\partial_1\phi)^2\right] \\ &+ \frac{1}{8}\left(\frac{dW}{d\phi}\right)^2 + \frac{1}{4}\bar{\psi}\psi\frac{d^2W}{d\phi^2}.\end{aligned} \quad (13.21)$$

13.2.3 Supersymmetry

Let us now consider the construction of the conserved supercurrent. From the supersymmetry transformations as in equations (13.10), we obtain that

$$\bar{\xi}J^m\Big|_{\text{symmetry}}$$

$$= \delta_\xi\phi\frac{\partial L}{\partial(\partial_m\phi)} + \delta_\xi\psi_\alpha\frac{\partial L}{\partial(\partial_m\psi_\alpha)} \quad (13.22a)$$

so that we get

$$J^m\Big|_{\text{symmetry}}$$
$$= \psi\partial^m\phi + \tfrac{1}{2}(\gamma^n\gamma^m(\partial_n\phi) - id\gamma^m)\psi. \qquad (13.22\text{b})$$

However, from equation (13.22a) we have

$$\partial_m\left\{\bar{\xi}J^m\Big|_{\text{symmetry}}\right\} = \delta_\xi L$$

$$= -\tfrac{i}{2}\bar{\xi}\gamma^m\partial_m\psi^{(\mathcal{L})}. \qquad (13.23)$$

In writing the last equation above, we have utilised that as stated explicitly in equation (13.17), L is a component field of the superfield $\mathcal{L}$ and thus transforms as per equation (13.10c) with (13.8). From equation (13.23) we easily obtain that the conserved supersymmetric current J^m is given as

$$J^m = J^m\Big|_{\text{symmetry}} + \frac{i}{2}\gamma^m\psi^{(\mathcal{L})}. \qquad (13.24)$$

From equations (13.14b) and (13.16), we have

$$\psi^{(\mathcal{L})} = (i\gamma^m\psi\partial_m\phi + d\psi) + \psi\frac{dW}{d\phi}. \qquad (13.25)$$

so that from equations (13.23) and (13.24), the conserved supercurrent is given as

$$J^m = \psi\partial^m\phi + \sigma^{nm}(\partial_n\phi)\psi - id\gamma^m\psi. \qquad (13.26)$$

In the above we have substituted

$$\sigma^{nm} = \frac{1}{2}(\gamma^n\gamma^m - \gamma^m\gamma^n). \qquad (13.27)$$

With a little algebra we can also explicitly verify that with J^m given by equation (13.26),

$$\partial_m J^m = 0, \qquad (13.28)$$

where we use field equations (13.19).

From equation (13.26) let us now consider the conserved supersymmetric charge given as

$$Q_\alpha = \int dx\, J^0{}_\alpha$$
$$= \int dx\,[\psi_\alpha(\partial_0\phi) - (\gamma^0\gamma^1\psi)_\alpha(\partial_1\phi) - id(\gamma^0\psi)_\alpha]. \qquad (13.29)$$

We then consider the unitary operator $U(\xi) = exp(i\bar{\xi}_\alpha Q_\alpha)$ and define the supersymmetric transformations of the field operators in Hilbert space e.g. through equations

13.2. Supersymmetry in 1+1 dimensions

$$U\phi(x)U^{-1} = \phi(x) + \delta_\xi \phi(x). \tag{13.30}$$

As usual, in the above x^0 in the definition of U and x^0 in ϕ are equal. Using quantisation noted subsequently, we may then easily verify that

$$\delta_\xi \phi \equiv [i\bar{\xi}Q, \phi(x)] = \bar{\xi}\psi(x). \tag{13.31a}$$

Similarly from equation (13.29) and with a little algebra we also obtain that

$$\delta_\xi \psi \equiv [i\bar{\xi}Q, \psi(x)] = -i\gamma^m \xi(\partial_m \phi) + d\xi, \tag{13.31b}$$

and that

$$\delta_\xi d \equiv [i\bar{\xi}Q, d] = -i\bar{\xi}(\gamma^m \partial_m \psi). \tag{13.31c}$$

In deriving the above equations, we have used the equal time quantisation conditions

$$[\phi(x), \partial_0 \phi(y)] = i\delta_1(x-y), \tag{13.32a}$$

$$[\psi_\alpha(x), \psi_\beta(y)]_+ = \delta_{\alpha\beta}\delta_1(x-y), \tag{13.32b}$$

Since the fields in equation (13.32b) are fermionic, we may regard the Dirac δ-function to be unusual, and may consider it e.g. as the limit given by

$$\delta(x) = \frac{1}{\sqrt{\pi}(a-b)}\left[exp\left(-\frac{x^2}{a^2}\right) - exp\left(-\frac{x^2}{b^2}\right)\right],$$

where $a > b > 0$ and $a \to 0$. This will ensure that the square of the fermionic operator in equation (13.32) at the same space point vanishes. We may note that equations (13.31) in the Hilbert space are the same as equations (13.10).

We shall now consider the algebra parallel to equation (13.6). We first note that the energy and momentum densities are given as

$$T^{00} = \frac{1}{2}\Big[-i\bar{\psi}\gamma^1\partial_1\psi + (\partial_0\phi)^2 + (\partial_1\phi)^2\Big]$$
$$+ \frac{1}{4}\left(\frac{dW}{d\phi}\right)^2 + \frac{1}{2}\bar{\psi}\psi\frac{d^2W}{d\phi^2}. \tag{13.33a}$$

$$T^{10} = -\frac{i}{2}(\partial^1 \psi_\alpha)\psi_\alpha$$
$$+ \frac{1}{2}\Big[(\partial^1\phi)(\partial^0\phi)+(\partial^0\phi)(\partial^1\phi)\Big]. \tag{13.33b}$$

The energy momentum four vector is given as

$$P^m = \int T^{m0} dx. \tag{13.34}$$

With supersymmetric charges Q_α given by equation (13.29), parallel to equation (13.6), we can now easily verify that

$$[Q_\alpha, Q_\beta]_+ = 2(\gamma^m \gamma^0)_{\alpha\beta} P_m. \tag{13.35}$$

In order to prove the above, we use equations (13.33) and (13.34) along with (13.32) for quantisation, as well as the identity

$$2[AB, CD]_+ = [A,C]_+[B,D]_+ + [A,C]_-[B,D]_-, \tag{13.36}$$

where, A and C are fermionic fields and B and D are bosonic fields in the integral in equation (13.29) with these fields commuting with each other. Calculations are particularly simple when we recognise that from equation (13.32a), $(\partial_0 \phi(x)) \equiv -i\delta/\delta\phi(x)$.

13.2.4 Field Expansion

We shall now consider an expansion of the bosonic and fermionic field operators to explicitly construct the Hilbert space. For the bosonic fields $\phi(x)$ and $\dot\phi(x)$ with $t=0$, we use the expansions,

$$\phi(x) = \frac{1}{\sqrt{2\omega_x}}\left[a(x) + a(x)^\dagger\right], \tag{13.37a}$$

$$\dot\phi(x) = i\sqrt{\frac{\omega_x}{2}}\left[a(x)^\dagger - a(x)\right], \tag{13.37b}$$

where, the annihilation and creation operators a and $a^\dagger$ satisfy the algebra $[a(x), a(y)^\dagger] = \delta(x-y)$. ω_x is an arbitrary differentiation operator and the form of equations (13.37) is such that the commutator algebra of equation (13.32a) is independent of ω_x. For the free fields however, we have explicitly

$$\omega_x = \left(-\left(\frac{\partial}{\partial x}\right)^2 + m_B^2\right)^{1/2}, \tag{13.38}$$

with m_B as the mass of the boson, but, for interacting fields it is basically unkonwn. The expansion of equation (13.37) is conventionally written in momentum space, but the above notation in coordinate space should be obvious when we note e.g. that for the annihilation operators

$$a(x) = \frac{1}{\sqrt{2\pi}} \int dk\, a(k) exp(ikx).$$

The anticommutator algebra of Majorana spinors in two dimensions as in equation (13.32b) is straightforward if we use Dirac method of quantisation or go through any modification of fermionic quantisation where canonically conjugate momenta become ill-defined [1].

Here we shall expand the fermionic fields rather in an unconventional manner as

$$\begin{aligned}\psi_\alpha &= \int \frac{1}{4\pi E}\Big[U_\alpha(k)c(k)\exp(ikx - iEt) \\ &+ V_\alpha(k)c(k)^\dagger \exp(-ikx + iEt)\Big]dk.\end{aligned} \quad (13.39)$$

In the above, for free fields $E = (k^2 + m_F{}^2)^{1/2}$ whereas for interacting fields it could be arbitrary. With the only nonvanishing anticommutators as

$$[c(k), c(k')^\dagger]_+ = 4\pi E \delta(k - k'), \quad (13.40)$$

we then obtain the anticommutators as in equation (13.32b) provided

$$U_\alpha(k)^\dagger U_\beta(k) + V_\alpha(-k)^\dagger V_\beta(-k) = 2E\delta_{\alpha\beta}, \quad (13.41)$$

For Majorana fields with equation (13.2) we clearly have $U(k) = V(k)^*$. For the free Dirac equation we in fact may take

$$U(k) = \begin{pmatrix} (E+k)^{\frac{1}{2}} \\ i(E-k)^{\frac{1}{2}} \end{pmatrix}, \quad (13.42)$$

We have in fact the invariant normalisations $\bar{U}U = 1 = -\bar{V}V$.

13.3 Quantum Phase Transition

We shall here illustrate with an example a simple mechanism of symmetry breaking as an example of destabilisation of vacuum as stated earlier in section 6.6 as well as give the temperature dependance for the same. For this purpose we explictly take

$$\frac{dW(\phi)}{d\phi} = \sqrt{(2\lambda)}(\phi^2 - m^2/\lambda), \quad (13.43)$$

so that the Lagrangian of equation (13.20) becomes

$$L = \frac{1}{2}\Big[i\bar\psi\gamma^m\partial_m\psi + (\partial_m\phi)(\partial^m\phi)$$
$$-\frac{\lambda}{2}\phi^4 + m^2\phi^2 - \frac{m^4}{2\lambda} - \sqrt{2\lambda}\phi\bar\psi\psi\Big]. \qquad (13.44)$$

This Lagrangian has a Z_2 symmetry as well as supersymmetry. We shall as stated tackle the problem of symmetry breaking with a variational method.

13.3.1 Breaking of Supersymmetry

Whether supersymmetry is broken or not can be checked by the action of supersymmetric generators Q_α on vacuum. Thus supersymetry is broken if $Q_\alpha \mid vac > \neq 0$; otherwise, it is unbroken. Hence, from the supersymmetry algebra of equation (13.35), $< vac \mid H \mid vac > \neq 0$ is equivalent to the breaking of supersymmetry. Thus we can regard this expectation value as an order parameter for supersymmetry breaking.

We further note that when $< vac \mid d \mid vac > \neq 0$, supersymmetry transformation (13.31b) can no longer be valid since vacuum expectation values of fermion fields can not be nonzero. Hence the vacuum expectation value of the auxiliary field may also be regarded as an order parameter for supersymmetry breaking.

In the model under consideration there is only one boson and one fermion. If their masses are different, then supersymmetry is broken. However, as we shall see, this is a sufficient condition, but is not a necessary condition.

13.3.2 Symmetry Breaking at zero temperature

We shall now consider symmetry breaking at zero temperature, which is conventional field theory. Let us define a new trial state for vacuum as

$$\mid vac' > = U(\xi) \mid vac >, \qquad (13.45a)$$

with a coherent state type of construction [2] for $U(\xi)$ given as

$$U(\xi) = \exp\Big[\xi\int dz(\omega_z/2)^{1/2}(a(z)^\dagger - a(z))\Big]. \qquad (13.45b)$$

In the above ξ is a real parameter and should not be confused with the earlier Majorana spinors for supersymmetry transformations. Field operator expansion (13.37) shall be used with $|vac>$ being characterised by annihilation operators in these expansions annihilating $|vac>$. The

13.3. Quantum Phase Transition

methodology here may be seen to be a continuation of what was done in section 6.6 for the consideration of symmetry breaking. With equations (13.45) and some minor algebra, we then see that

$$< vac' \mid \phi \mid vac' > = \xi, \qquad (13.46a)$$

and that

$$< vac' \mid \phi' \mid vac' > = 0, \qquad (13.46b)$$

where

$$\phi' \equiv U(\xi)\phi U(\xi)^{-1} = \phi - \xi. \qquad (13.46c)$$

From equation (13.21) with $\frac{dW}{d\phi}$ as in (13.43) we then easily calculate that

$$V(\xi) \equiv < vac' \mid H \mid vac' >$$

$$= \frac{\lambda}{4}\xi^4 - \frac{m^2}{2}\xi^2 + \frac{m^4}{4\lambda}. \qquad (13.47)$$

The parameter ξ is determined by minimising the expectation value of the energy density. Thus

$$\xi = \xi_{min} = \left(\frac{m^2}{\lambda}\right)^{1/2}. \qquad (13.48)$$

We may identify the mass square of the boson as given by

$$m_B{}^2 = \frac{d^2V}{d\xi^2}\bigg|_{\xi=\xi_{min}}, \qquad (13.49)$$

and mass of the fermion given as the appropriate coefficient of $\bar{\psi}\psi$. Then, we easily identify that

$$m_B = m_F = \sqrt{2}m. \qquad (13.50)$$

Also, we easily see that

$$< vac' \mid d \mid vac' > = 0, \qquad (13.51)$$

so that, as stated in subsection 13.3.1, supersymmetry remains unbroken. However, equation (13.46c) implies that the effective Lagrangian shall no longer have Z_2 symmetry, which is spontaneously broken.

We have thus shown that breaking of Z_2 symmetry can be regarded as a restructuring of vacuum state, with $\mid vac' >$ for broken phase being expressed in terms of perturbative $\mid vac >$ through a unitary transformation, which is parallel to section 6.6. This restructuring does not break supersymmetry. We shall next discuss phase transition at finite temperature, again as a restructuring of the "vacuum state" which becomes temperature dependant, using the thermofield method of Umezawa [3] as well as some self-consistency requirements.

13.4 Temperature dependance of phase transition

Supersymmetry at finite temperature in 1+1 dimensions was first considered by Das and Kaku [4]. Using thermofield method it has also been considered in Ref. [5] in four dimensions. This usually involves a high temperature approximation. We shall here in contrast look into this by a variational method as vacuum realignment parallel to section 6.6 which gives some additional insight.

When we have a physical system at some temperature, this system is described by a density operator given as $exp(-\beta H)$ where β is the inverse of temperature. However, at zero temperature we have for any operator O

$$lim_{\beta \to \infty} \frac{Tr[exp(-\beta H)O]}{Tr[exp(-\beta H)]} = \frac{<vac' \mid O \mid vac'>}{<vac' \mid vac'>}, \qquad (13.52)$$

with $\mid vac' >$ being the state of lowest energy. Thus, the thermodynamic expectation value becomes here a quantum expectation value. However, at finite temperature the above identity does not hold. We do not e.g. get the fermi or bose distribution functions when we take the expectation values for the corresponding number density operators. Thermofield method [3] takes care of this by doubling the Hilbert space, so that the statistical (thermal) average becomes a quantum expectation value as in the right hand side of equation (13.52) in the extended Hilbert space yielding the appropriate fermi or bose distribution functions.

Thus we introduce new bosonic and fermionic annihilation operators $\tilde{a}(k)$ and $\tilde{c}(k)$ parallel to those already discussed in the last section. We then define the temperature dependant ground state $\mid vac', \beta >$ using the above "thermal" modes through an unitary transformation parallel to what was done in the last section. Explicitly, we take

$$\mid vac', \beta >= U_B(\beta) U_F(\beta) \mid vac' >, \qquad (13.53)$$

with

$$U_B(\beta) = exp\left[\int dk g(k,\beta)\Big(a'(k)^\dagger \tilde{a}'(-k)^\dagger - \tilde{a}'(-k)a'(k)\Big)\right], \qquad (13.54a)$$

$$U_F(\beta) = exp\left[\int dk f(k,\beta)\Big(c(k)^\dagger \tilde{c}(-k)^\dagger - \tilde{c}(-k)c(k)\Big)\right], \qquad (13.54b)$$

13.4. Temperature dependance of phase transition

In the above, $a'(k)$ and $a'(k)^\dagger$ are annihilation and creation operators corresponing to the field ϕ' of the last section defined through equation (13.46c). $g(k,\beta)$ and $f(k,\beta)$ will be subsequently defined so as to obtain the correct distribution functions for the bosons and the fermions.

Corresponding to the unitary transformation of the vacuum state, we now define the temperature dependant boson annihilation operators which annihilate the thermal vacuum through

$$a'(k,\beta) = U_B(\beta) a'(k) U_B(\beta)^{-1}, \qquad (13.55a)$$

and

$$\tilde{a}'(k,\beta) = U_B(\beta) \tilde{a}'(k) U_B(\beta)^{-1}. \qquad (13.55b)$$

From the above equations, we can see that the temperature independent bosonic operators are related to the temperature dependant bosonic operators through the Bogoliubov transformation

$$\begin{pmatrix} a'(k) \\ \tilde{a}'(-k)^\dagger \end{pmatrix} = \begin{pmatrix} \cosh(g(k,\beta)) & \sinh(g(k,\beta)) \\ \sinh(g(k,\beta)) & \cosh(g(k,\beta)) \end{pmatrix} \begin{pmatrix} a'(k,\beta) \\ \tilde{a}'(-k,\beta)^\dagger \end{pmatrix}. \qquad (13.56)$$

Similarly, for the fermionic sector the operators which annihilate the thermal vacuum are given as

$$c(k,\beta) = U_F(\beta) c(k) U_F(\beta)^{-1}, \qquad (13.57a)$$

and

$$\tilde{c}(k,\beta) = U_F(\beta) \tilde{c}(k) U_F(\beta)^{-1}. \qquad (13.57b)$$

As earlier we can now see that the temperature independent operators are related to the temperature dependant operators through the Bogoliubov transformation

$$\begin{pmatrix} c(k) \\ \tilde{c}(-k)^\dagger \end{pmatrix} = \begin{pmatrix} \cos(f(k,\beta)) & \sin(f(k,\beta)) \\ -\sin(f(k,\beta)) & \cos(f(k,\beta)) \end{pmatrix} \begin{pmatrix} c(k,\beta) \\ \tilde{c}(-k,\beta)^\dagger \end{pmatrix}. \qquad (13.58)$$

Our next job is to evaluate the finite temperature correction to the effective potential given by equation (13.47). For this purpose we substitute ϕ in terms of ϕ' in equation (13.44) and collect the bosonic part of the Hamiltonian and the fermionic kinetic and interaction terms of the same to obtain respectively

$$H_B = \frac{1}{2}((\partial_0\phi')^2 + (\partial_1\phi')^2)$$
$$+ \frac{\lambda}{4}\phi'^4 + \lambda\xi\phi'^3 + \frac{1}{2}(3\lambda\xi^2 - m^2)\phi'^2$$
$$+ (\lambda\xi^3 - m^2\xi)\phi' + \frac{\lambda}{4}\xi^4 - \frac{m^2\xi^2}{2} + \frac{m^4}{4\lambda}, \qquad (13.59a)$$

and

$$H_F + H_{\text{int}} = -\frac{i}{2}\bar\psi\gamma^1\partial_1\psi + \left(\frac{\lambda}{2}\right)^{1/2}[\phi'\bar\psi\psi + \xi\bar\psi\psi]. \qquad (13.59b)$$

For the evaluation of the expectation value of the bosonic part corresponding to $|vac',\beta>$, we note that from equation (13.56) we have

$$<vac',\beta|a'(k)a'(k')|vac',\beta> = 0, \qquad (13.60a)$$

$$<vac',\beta|a'(k)^\dagger a'(k')^\dagger|vac',\beta> = 0 \qquad (13.60b)$$

and

$$<vac',\beta|a'(k)^\dagger a'(k')|vac',\beta>$$
$$= \sinh^2 g(k,\beta)\delta(k-k'). \qquad (13.60c)$$

In order ensure that the appropriate bosonic distribution is obtained, or through an extremisation of free energy including entropy in thermofield dynamics, we derive that

$$sinh^2 g(k,\beta) = \frac{1}{exp(\beta\omega(k,\beta)) - 1}, \qquad (13.61a)$$

where

$$\omega(k,\beta) = (k^2 + m_B(\beta)^2)^{1/2}, \qquad (13.61b)$$

with $m_B(\beta)$ as the effective boson mass at finite temperature to be subsequently defined in a self-consistent manner. Similarly for the expectation values of the fermionic operators at finite temperature, we note that from equation (13.58) we have,

$$<vac',\beta|c(k)c(k')|vac',\beta> = 0, \qquad (13.62a)$$

$$<vac',\beta|c(k)^\dagger c(k')^\dagger|vac',\beta> = 0, \qquad (13.62b)$$

and

13.4. Temperature dependance of phase transition

$$< vac', \beta \mid c(k)^\dagger c(k') \mid vac', \beta > = \sin^2 f(k,\beta) 4\pi E \delta(k-k'), \tag{13.62c}$$

Again, parallel to equations (13.61), here we have

$$\sin^2 f(k,\beta) = \frac{1}{\exp{(\beta\epsilon(k,\beta))} + 1}, \tag{13.63a}$$

where

$$\epsilon(k,\beta) = (k^2 + m_F(\beta)^2)^{1/2}, \tag{13.63b}$$

with $m_F(\beta)$ as the effective mass of the fermion at finite temperature.

In order to obtain the temperature dependant effective potential $V(\xi,\beta)$, we now take the expectation value of the Hamiltonian of equation (13.59), and parametrise the expansion of the field operators in terms of the effective masses of the fermions and bosons. With equations (13.60) to (13.63) for the expectation values as needed, we then obtain that

$$\begin{aligned} V(\xi,\beta) &= V_1(\xi,\beta) + V_2(\xi,\beta) + \frac{\lambda}{4}\left(\xi^2 - \frac{m^2}{\lambda}\right)^2 \\ &\quad + V_3(\xi,\beta) + V_4(\xi,\beta), \end{aligned} \tag{13.64}$$

where the contributions V_i with $i = 1, \cdots, 4$ are given as

$$V_1(\xi,\beta) = \frac{1}{4\pi} \int_{-\infty}^{\infty} \frac{(3\lambda\xi^2 - m^2 + k^2 + \omega(k,\beta)^2)}{\omega(k,\beta)(e^{\beta\omega(k,\beta)} - 1)} dk, \tag{13.65a}$$

$$V_2(\xi,\beta) = \frac{3\lambda}{4}\left[\frac{1}{2\pi}\int_{-\infty}^{\infty}\frac{dk}{\omega(k,\beta)(e^{\beta\omega(k,\beta)}-1)}\right]^2, \tag{13.65b}$$

$$V_3(\xi,\beta) = \frac{1}{2\pi}\int_{-\infty}^{\infty}\frac{k^2 dk}{\epsilon(k,\beta)(e^{\beta\epsilon(k,\beta)}+1)}, \tag{13.65c}$$

and

$$V_4(\xi,\beta) = \sqrt{\frac{\lambda}{2}}\cdot\frac{1}{\pi}\cdot\xi\cdot m_F(\beta)\cdot\int_{-\infty}^{\infty}\frac{dk}{\epsilon(k,\beta)(e^{\beta\epsilon(k,\beta)}+1)}. \tag{13.65d}$$

For convenience of calculations we shall now write the above expressions in terms of dimensionless variables in terms of the mass parameter m. Thus we substitute $\lambda' = \lambda/m^2$, $\mu_B(\beta) = m_B(\beta)/m$, $\mu_F(\beta) = m_F(\beta)/m$, $y = \beta m$, and choose the integration variable x=k/m. We then obtain,

$$V(\xi,\beta) = m^2\Big[\frac{1}{2}I_1(y) + \frac{3\lambda'}{4}I_2(y)^2 + \frac{\lambda'}{4}(\xi^2 - \frac{m^2}{\lambda'})^2$$
$$+ I_3(y) + I_4(y)\Big], \qquad (13.66)$$

where, with integrations in x from zero to infinity,

$$I_1(y) = \frac{1}{\pi}\int_0^\infty \frac{(3\lambda'\xi^2 - 1 + x^2 + \omega(x,y)^2)}{\omega(x,y)(e^{y\omega(x,y)} - 1)}dx, \qquad (13.67a)$$

$$I_2(y) = \frac{1}{\pi}\int_0^\infty \frac{dx}{\omega(x,y)(e^{y\omega(x,y)} - 1)}, \qquad (13.67b)$$

$$I_3(y) = \frac{1}{\pi}\int_0^\infty \frac{x^2 dx}{\epsilon(x,y)(e^{y\epsilon(x,y)} + 1)} \qquad (13.67c)$$

and

$$I_4(y) = \sqrt{2\lambda'}\cdot\frac{1}{\pi}\cdot\xi\cdot\mu_F(y)\cdot\int_0^\infty \frac{dx}{\epsilon(x,y)(e^{y\epsilon(x,y)} + 1)}. \qquad (13.67d)$$

In the above for bosons and fermions we have substituted

$$\omega(x,y) = (x^2 + \mu_B(y)^2)^{1/2}, \qquad (13.68a)$$

and

$$\epsilon(x,y) = (x^2 + \mu_F(y)^2)^{1/2}. \qquad (13.68b)$$

We first note that for the temperature dependant effective potential corresponding to $|vac', \beta >$, at zero temperature $V_i(i = 1, \cdots, 4)$ are zero, so that then equation (13.64) becomes trivially the same as in equation (13.47). Thus, the above four contributions are the corrections to the potential at finite temperature. When the potential is known, we minimise the same to determine $\xi = \xi_{min}(\beta)$, naturally as a function of temperature.

The above calculations, however, need a self-consistent determination of the masses of the fermions and the bosons. Firstly, at any temperature, the evaluation of the integrals in equation (13.66) needs the

13.4. Temperature dependance of phase transition

above effective masses as inputs. When the potential is minimised and $\xi_{min}(\beta)$ is determined, the fermion mass can be obtained from equation (13.59b) as $(2\lambda)^{1/2}\xi_{min}(\beta)$, and the boson mass can be determined from equation (13.49) again as function of temperature. Self-consistency requires that the input masses in the integrals for the determination of the potential shall be the same as the output masses obtained after the potential is minimised. At any given temperature, this needs an iterative calculation for the masses of the fermions and the bosons. When the masses are consistently determined, the corresponding potential becomes acceptable [6].

With numerical computations we can see that here at any finite temperature, supersymmetry gets broken, since the vacuum energy density becomes positive definite, and the auxiliary field also attains a nonzero expectation value [6]. Further, the fermion and boson masses remain degenerate which e.g. is similar to the case considered in Ref.[5] until a critical temperature $T_c \simeq 2m$ is reached. For $T > T_c$, Z_2 symmetry is restored, supersymmetry remains broken, and fermion becomes massless lifting the fermion boson mass degeneracy present below T_c.

We shall consider a simple version of Goldstone theorem for the symmetry breaking as above. With the supersymmetric charge defined as in equation (13.29), for the Hamiltonian H, $[Q_\alpha, H] = 0$ which reflects the fact that Q_α is conserved. For supersymmetry breaking as above, we also have

$$Q_\alpha \mid vac'; \beta > \neq 0. \qquad (13.69)$$

Clearly $Q_\alpha \mid vac'; \beta >$ creates a state of zero total momentum. Furthermore, $[Q_\alpha, H] = 0$ implies that for any eigenstate $\mid h >$ of H, $Q_\alpha \mid h >$ has the same eigenvalue h of H. Hence Q_α must create zero energy modes. From the Bogoliubov transformations (13.56) we can see that at finite temperatures Q_α create a composite super pairs consisting of a fermion and a thermal boson or a thermal fermion and a boson, of zero total momentum. We note that for $T < T_c$ these are the Goldstino modes [5]. With thermofield dynamics it can be recognised as such when we note that here the total hamiltonian with thermal modes consists of the conventional Hamiltonian minus the thermal hamiltonian [3], and the fermion and boson masses are degenerate. The existence of the corresponding singularity in the temperature dependant Greens function has been examined in Ref [5]. However, when $T > T_c$, masses of fermions and bosons are different, and thus the above mechanism does not operate to give a Goldstone mode. But, the mass of the fermion now becomes zero, and thus this fermion is the Goldstino for supersymmetry breaking, which is a more conventional picture for Goldstone theorem for the breaking of supersymmetry as discussed e.g. in section 6.3.

13.5 Discussions

Our objective here has been to give an example of breaking of supersymmetry beyond the tree level by a nonperturbative variational method, where symmetry breaking takes place with a restructuring of the vacuum state through a unitary transformation. This methodology could be illustrated with relative ease in 1+1 dimensions since with the example we choose, we have only one boson and one fermion field. It becomes computationally more prohibitive when larger number of fields are involved. However, it gives fresh insight into symmetry breaking [7] including possible experimental signatures for phase transitions.

REFERENCES

1. S.P. Misra and T. Pattnaik, Phys. Lett. 129B, 401 (1983); J. Barcelos-Neto and A. Das, Phys. Rev. D33, 2863 (1986).

2. S.P. Misra, Phys. Rev. D35, 2607 (1987); H. Mishra, S.P. Misra and A. Mishra, Int. J. Mod. Phys. A3, 2311 (1988).

3. H. Umezawa, H. Matsumoto and M. Tachiki, *Thermofield dynamics and condensed states* (North Holland, Amsterdam, 1983).

4. A. Das and M. Kaku, Phys. Rev. D18, 4540 (1978).

5. H. Matsumoto, M. Nakahara, Y. Nakano and H. Umezawa, Phys. Rev. D29, 2838 (1984).

6. A. Mishra, H. Mishra, S.P. Misra and S.N. Nayak, Phys. Lett. B251, 541 (1990).

7. S.P. Misra, *Proceedings of Workshop on High Energy Phenomenology*, edited by D.P. Roy and P. Roy (World Scientific, Singapore, 1989) p. 346.

Chapter 14

Conclusions

In the text we have seen the varied nature of the theory as well as applications of supersymmetry and supergravity. It is obvious that this constitutes a significant extension of our ideas of symmetry, since here internal symmetry gets related to Lorentz symmetry in a more intimate manner. Further, we no longer deal with a Lie algebra corresponding to a continuous group, but have instead a graded Lie algebra with both commutators as well as anticommutators of the generators. This in itself goes around some no go theorems [1] regarding Lorentz group structure being subsumed in a larger mathematical (physical) space. One would naturally ask now whether we have any positive experimental signature for the same. One may also ask the question that all our schemes have dealt with quarks and leptons being elementary objects even upto Planck scale, or, the compactification scale. On the other hand, our experience in physics is that as we probe into smaller and smaller distances (or, with larger and larger energies), we come across fresh degrees of compositeness. We have so far ignored any such question. It shall be useful to have a picture of some such possible scenario. Further, conformal field theory and some other related developments have some attractive features as a description of physics related to string theories. The nice feature of such models is that most of dynamics gets converted to symmetry, and thus becomes almost solvable. We have stated some results of the same in Chapter 12. We shall make some further observations regarding the same in the context of matrix models.

14.1 Experimental Signatures

One of the basic experimental observations one would like to make is to directly see the supersymmetric partners of the low energy particles, such as, photinos, scalar leptons or scalar quarks, or gluinos. The phenomenology for the interactions as well as the production of supersymmetric particles is described in some detail by Haber and Kane [2] where the Lagrangians mentioned in Chapter 9 earlier are used with specific symmetry groups. The general scenario of such productions is known. The principle which is employed for the observations of a signal for production of supersymmetric particles as such is R-parity, described in subsection 6.7d, which is regarded as conserved. R-parity as explained earlier is a multiplicative quantum number, which is odd for the supersymmetric parnters of the known particles, and is even for the original familiar particles. The conservation of R-parity leads to the conclusion that the lightest supersymmetric particle (LSP) should be stable. Hence, in high energy collisions, like the neutrino, LSP will generally interact weakly and escape detection, leading to a momentum imbalance in observed jets and hadrons. There will also be an associated signal which will depend on the mode of its production and naturally on which supersymmetric partner is the lightest. One expects this to be either the photino, or the scalar neutrino. No such missing energy or momentum beyond the standard model has been yet seen. At present the experimental lower bounds for the masses of scalar quarks is about 100 GeV and for gluinos it is more than the above because being color octets they interact more strongly, so that signal to noise ratio for their production for a given threshold is higher. The scalar leptons shall be produced in e^+e^- colliders, and thus the limit for them is around 40 GeV. As stated, so far no missing momentum has been identified which can not be accounted by the neutrinos of the known light particle spectrum, so that the lightest supersymmetric particle with momentum or energy imbalance is not yet seen.

The scenario of exotic particle associated with supersymmetric partners of known particles in the context of larger symmetry groups like E_6 is a little richer. We may discuss the production of the same in the context of string based supersymmetry. Let us consider the particle content in either $SU(5)$, $SO(10)$ or E_6 theories. As given in Table 3 of Appendix B, we can write the 27-plet of E_6 in tems of $SO(10)$ and $SU(5)$ multiplet structures as

$$27 = 16 + 10 + 1 = (10 + 5^* + 1) + (5 + 5^*) + 1 \qquad (14.1)$$

Thus, in E_6 type of supergravity expected from superstring inspired phenomenology, we shall have other exotic light particles around masses 100 GeV to 1 TeV in addition to fifteen quarks and leptons of a single generation. None of these particles so far have been observed [3]. However, this may be merely due to fact that experimentally we are now just reaching such energy scales, and observation of such particles being round the corner is quite possible. We are thus at present neither able to assert that supersymmetry is there on experimental grounds, nor in a position to reject such a possibility. As often happens in physics, and as really happened for the discovery of electroweak symmetry, our main consideration here therefore has been aesthetic appeal. We may particularly note that the natural energy scale for supergravity is Planck scale, which is about 10^{19} GeV. A theoretical insight may give us a sense of direction for further developments as well as an understanding of nature of dynamics at such high energy scales. At present we do not have this. We shall now briefly recapitulate some other ideas which have been proposed in the context of supersymmetry and supergravity, which could give us the direction for future developments.

14.2 Preonic Models

As mentioned, it has been our experience in physics that as we go to lower distance scales, or higher energies, we encounter fresh degrees of compositeness of matter, as well as discover fresh dynamical laws. It shall therefore be presumptuous on our part to imagine that at present we have discovered the ultimate structure of matter at small distances. It is then natural to ask ourselves whether in fact quarks and leptons, which so far are the 'final' building blocks for our dynamics, are themselves composite.

There is a conceptual difference between quarks and leptons being composite objects and the other composite states in quantum field theory known at present. From experiments, the size r_0 of leptons is determined to be smaller than 10^{-16} cms. Thus the scale $\Lambda_M = 1/r_0$ for their inverse size is greater than a few hundred GeV. However, their masses are much smaller than a GeV. If quarks and leptons are composite objects, besides other features we should have an understanding of this anomaly.

For quarks and leptons to be composite, we would like to have the following: (i) the scale Λ_M at which quarks and leptons are composite should be much larger than the weak scale, the reason for this being that we do not find any unexpected phenomena indicating finite size upto a scale higher than the weak scale. (ii) Also, the inverse size should be smaller than the Planck scale, since as per our present prejudices, at Planck scale we have superstring theories some variation of which could be the "ultimate" theory. (iii) The weak scale of about 100 GeV should occur naturally within the scheme. (iv) The masses of the quarks and leptons should be protected so as not to occur either at scale Λ_M or at the weak scale, but should occur at a still lower scale. In addition (v) we would like to have an understanding of the quantum numbers of quarks and leptons at a fundamental level, the replication of generations including the generation of the masses and their mixings, in case it is possible to get these.

There have been many attempts regarding composite models. We shall here however illustrate with one particular model as an illustration [4], which is relatively simple and where some ingenious ideas of supersymmetry have been used [5]. We shall particularly examine the role of supersymmetry and supergravity concepts in the development of these ideas.

Let us first consider the quantum numbers of quarks and leptons. We assume that these quantum numbers arise as a signal of some 'elementary' objects which carry these quantum numbers. Let each of these quantum numbers be generated by a chiral superfield Φ, the number of such chiral superfields being n_p. We call these as preons. Thus, if we have n_f flavors, and n_c colors, then

$$n_f + n_c = n_p. \qquad (14.2)$$

Let us consider a single generation. We then have two flavors, and, if we take the leptons degree of freedom to indicate the fourth color, we have four colors. In this case we shall have $n_f = 2$ and $n_c = 4$, such that $n_p = 6$. In flavor space, we have noticed since a long time that e.g. $SU(2)_L \times SU(2)_R$ global symmetry through current algebra [2] gives useful results. Parallel to the same we take the global symmetry [4].

$$SU(n_p)_L \times SU(n_p)_R \times U(1)_V \times U(1)_X. \qquad (14.3)$$

In the above, besides the chiral global symmedtry, we have taken the extra symmetries $U(1)_V$ and $U(1)_X$ which respectively correspond to the conservation of preon number, and, nonanomalous R-symmetry. We recall that R-symmetry is useful to forbid rapid proton decay, and is utilised to identify signals for the production of supersymmetric particles.

Let us next consider the origin of scale larger than the weak scale. We are at present aware of a new type of origin of such scales from the analysis of renormalisation group equations [8], as in equation B.11 of appendix B. This has been used repeatedly in grand unification theories. We know here that $SU(N)$ symmetries have asymptotic freedom, and when we consider a decrease in energy scale, the running coupling constant increases giving rise to confinement. Such a statement is not strictly logical, but is an intuitive way of understanding confinement. Thus, e.g. quantum chromodynamics is associated with a scale arising dynamically, which is the QCD parameter $\Lambda \cong 200$ MeV. We presume that there is a metacolor group $SU(N)$ where confinement occurs at a scale of $\Lambda_M \sim 10^{12}$ GeV. This will correspond to the size parameter of

14.2. Preonic Models

the composite particles, believed to be elementary at the present scale. We subsume this gauge symmetry with local supersymmetry or supergravity as utilised in the earlier Chapter. We thus take a metacolor gauge group $SU(N)$ where it is called metacolor since the mechanism is similar to that of color. In Wess-Zumino gauge, we then have the gauge multiplet $V = (v_\mu, \lambda, D)$.

Let us next examine the possibility that such composite particles, arising from the dynamics at the scale of Λ_M shall have small masses as observed. We have here two scales - one is the Planck scale m_{Pl} as an inevitable scale when we consider supergravity, and the other is Λ_M as the confinement scale of metacolor gauge symmetry. The interplay of these scales should produce the scales as desired. We have earlier considered such an interplay of scales many times. For example, the mass scale of Polonyi potential and the Planck scale generated the weak scale as well as the scale of sypersymmetry breaking as seen in Section 10.2 and elsewhere. In Section 10.3.3, a similar interplay of scales gave rise to heavy as well as ultralight neutrinos. In all these cases vacuum expectation values of scalar fields yielded the different scales as mentioned. However, in Ref.[9] and as shown in Section 10.4, we have the alternative example of the scale Λ_M of fermion condensate generating the low energy scales Λ_M^3/m_{Pl}^2 or Λ_M^3/m_{GUT}^2 for the weak interactions. We shall take here a mechanism similar to the above, where using supersymmetry, we shall also obtain a special mechanism for the protection of masses of quarks and leptons [5] and some other useful results.

Let us consider the chiral multiplets Φ_+ and Φ_-^* belonging to conjugate representations of the metacolor group $SU(N)$ with component chiral multiplets being denoted by $(\phi_L^a, \psi_L^a, F_L^a)$ and $(\phi_R^a, \psi_R^a, F_R^a)^*$ respectively. We have in the above suppressed the metacolor group indices, and, a runs from 1 to n_p. For metacolor global supersymmetry we then have the Ward identities given as [10]

$$\frac{1}{\sqrt{2}} < vac \mid [Q, \bar{\psi}_R^a \, \phi_L^b]_+ \mid vac >$$

$$= < vac \mid F_R^{a*} \, \phi_L^b \mid vac > \; - \; < vac \mid \bar{\psi}_R^a \, \psi_L^b \mid vac > \qquad (14.4)$$

and

$$\frac{1}{2\sqrt{2}} < vac \mid [\bar{Q}, \bar{\psi}_R^a \, \phi_R^a]_+ \mid vac >$$

$$= < vac \mid F_R^{a*} \, \phi_R^a \mid vac > \; + \frac{g^2}{32\pi^2} < vac \mid \lambda \cdot \lambda \mid vac > . \qquad (14.5)$$

In the above, Q and $\bar{Q}$ are spinorial generators of supersymmetry, and we have considered vacuum expectation values. Hence it is clear that if the right hand side expressions do not vanish, then global supersymmetry shall be broken. We assume that when m_{Pl} approaches infinity, global supersymmetry gets restored, so that the vacuum expectation values of the condensates on the right hand side of equations (4) and (5) in this limit must vanish. Since such a limit should be smooth, for finite m_{Pl} we expect that the nonzero vacuum expectation values of the condensates must be supressed by factors of m_{Pl}. We consider such a limit keeping the scale for metacolor Λ_M fixed, and making m_{Pl} go to infinity. We thus expect that the vacuum expectation values of the condensates have the form

$$< vac \mid \lambda \cdot \lambda \mid vac > = a_\lambda \Lambda_M{}^3 ((\Lambda_M/m_{Pl})^{n_\lambda} \qquad (14.6)$$

and

$$< vac \mid \bar\psi_R^a \psi_L^b \mid vac > = a_\psi \Lambda_M{}^3 \, A_{ab} \left((\Lambda_M/m_{Pl})^{n_\psi}\right) \qquad (14.7)$$

where, we expect the constant n_λ and n_ψ to be positive, and shall take a_λ and a_ψ to be of the order of one. Such a result should be capable of being arrived at dynamically, but since the process basically is nonperturbative, we shall not try it.

The choice of quantum numbers n_p as corresponding to the color and flavor degrees of freedom indicates that the observed quarks and leptons shall be composites of the chiral superfields Φ_-^a and Φ_+^{b*}. We may take e.g.

$$(q_R^{ab*})^\dagger = (\psi_R^a \phi_L^{b*} + \phi_R^a \psi_L^{b*})^\dagger \qquad (14.8)$$

which creates color and flavor quantum numbers from two separate chiral superfields. We presume the mass term $\bar q_R q_L + \bar q_L q_R$ to come from both the condensates $\bar\psi_R \psi_L$ and $\bar\phi_R \phi_L$

Thus from equation (14.7) the quark masses will be supressed by the factor $(\Lambda_M/m_{Pl})^{n_\psi}$ compared to Λ_M. We note that these arguments are heuristic, and until we shall be able to perfect a reasobale approximation to deal with both the vacuum structure in such theories, as well as the same for the formation of composite objects with such exceptionaly complicated dynamics, more precise quantitative understanding will not be possible. With similar arguments, parallel to equations (14.6) and (14.7), we can also show that

$$< vac \mid \phi_R^{a*} \phi_L^b \mid vac > = a_\phi \Lambda_M{}^2 M_{ab} (\Lambda_M/m_{Pl})^{n_\phi}. \qquad (14.9)$$

In fact, the scalar condensate above is introduced perturbatively, so that a_ϕ is of the order of a_λ, a_ψ, and,

14.2. Preonic Models

$$n_\phi = n_\lambda + n_\psi \qquad (14.10)$$

The theory is so constructed that the mass scales for breaking of supersymmetry δm_s is around 1 TeV and the masses of W and Z bosons are around 100 GeV, leading to the hierarchy structure

$$m_W < \delta m_s \ll \Lambda_M \ll m_{Pl}. \qquad (14.11)$$

We shall now very briefly note the see-saw mechanism for the generation of the quark lepton masses. Let $q_R^{fc*} \sim \psi_R^f \phi_L^{c*} + \phi_R^f \psi_L^{c*}$, which is a part of $\Phi_-^f \Phi_+^{c*}$, and, similarly $q_L^{fc*} \sim \psi_L^f \phi_R^{c*} + \phi_L^f \psi_R^{c*}$, which is a part of $\Phi_+^f \Phi_-^{c*}$. In the above f and c carry the flavor and color quantum numbers of the quarks through the appropriate superfields, and leptons have the fourth color. Then the quark masses are generated through the product of

$$< vac \,|\, \bar{\psi}_R \psi_L \,|\, vac > \quad \text{and} \quad < vac \,|\, \phi_R^* \phi_L \,|\, vac > .$$

Hence, as stated above, the mass terms is given by

$$m^{(0)}(q_L \leftrightarrow q_R) \sim a_\psi a_\phi \Lambda_M (\Lambda_M/m_{Pl})^{n_\psi + n_\phi} \sim 1 MeV. \qquad (14.12)$$

We next note that we can also have other types of composites formed from $\Phi_+^f \Phi_+^{c*}$ and $\Phi_-^f \Phi_-^{c*}$ leading to metacolor singlets and these will have the same flavor and color quantum numbers as quarks and leptons. The construction is similar, except that since here we are having products of chiral and antichiral superfields, the resulting superfield will be reducible as far as supersymmetry is concerned with appropriate structures. We shall thus have fresh composites denoted by Q_L and Q'_L along with the corresponding right handed components. These shall be new vector like families with mass around 1 TeV region as a prediction of the scenario. The chiral families arise here due to the fermion boson pairing in supersymmetry. We note that in the above we always mean quarks and leptons when we say quarks, since leptons are taken to differ from quarks through a fourth color quantum number.

Next we include mixing and write down a general mass matrix which is symmetric under left-right symmetry. This block matrix in q, Q and Q' has a reasonable restricted form [4], which, through a see-saw mechanism generates the different mass scales for the individual generations [4]. One also assumes e.g. that $< vac \,|\, \lambda \cdot \lambda \,|\, vac >$ is larger than $< vac \,|\, \bar{\psi}\psi \,|\, vac >$ since the coupling for the adjoint representations is always stronger than that in the fundamental representation. One then obtains relations for the ratios of the masses of individual generations, in reasonable agreement with experiments. Also, CP violation can be incorporated with a mechanisms similar to that of left-right symmetric model [6] with the difference that here it takes place through condensates, and hence shall survive even when W_R become superheavy.

It may be clear that the results are qualitative in nature. We have here given a limited description of compositeness, and it may be recognised that quite a number of details have been omitted. It has been our intention to give some feeling as how many of the features of quarks and leptons can emerge as a result of compositeness. However, the arguments per force are heuristic. We would like to have some methodology for controlling the corresponding dynamics to make quantitative predictions which may be even approximately true. It is obvious that there are a number of dynamical questions which must be answered when we wish to examine the full content of compositeness. As per our previous experience, when higher scales are involved, we do not see any reason not to expect composite objects to be formed at that scale. Hence irrespective of whether quarks and leptons are composite or otherwise, we should have an understanding of the dynamics of formation of composite objects at high energy scales when supersymmetry is present. Only then we shall be able to extrapolate our ideas to such scales. It is clear that such dynamical questions should be resolved soon so that we may have confidence in extrapolating our ideas to such energy scales.

14.3 Matrix Models

Recently there has been a lot of interest in matrix models in the context of some nonperturbative solutions in string theory. String theory is a two dimensional field theory. In functional methods introduced in Chapter 7 one takes an integration over all functions defined in two dimensions. The two dimensional surface may have the global characteristic of having different topologies. We may imagine it for example to be a sphere, or a sphere with a single 'hole' on it, or one with two 'holes', and so on. The surface containing n 'holes' or 'handles' is said to have genus n. The continuous functions on each of these surfaces will lead to different functional integrals in the evaluation of the 'partition function' as appropriate for the specific dynamical system.

One attempts to solve such problems through a discretisation procedure, so that the continuous space is replaced by an infinite number of discrete points, and then modulates the dynamics accordingly to simulate the continuum dynamics. In a plane, the most familiar technique used by mathematicians since quite some time is through triangulation. We can do this maintaining the topological features of the surface under discussions; in fact, it has been an important method for discussing

14.3. Matrix Models

the nature of topology of a surface. Here we are interested in the same for the evaluation of the relevant functional integrals for the discussion of a physics content of a theory.

We shall be particularly interested with the application of the above ideas in the context of supersymmetry. We shall thus consider a string which is supersymmetric in target space [11]. For this purpose we replace the supercoordinate $F(t,\theta,\bar{\theta})$ of Chapter 2 by $N \times N$ matrix superfield Φ. Thus in superspace $(t,\theta,\bar{\theta})$ parallel to equation (2.1) we write the matrix supercoordinate as

$$\Phi = X + \theta\Psi + \bar{\theta}\bar{\Psi} + \theta\bar{\theta}A. \tag{14.13}$$

In the above X is the matrix bosonic coordinate, Ψ and $\bar{\Psi}$ are the matrix fermionic coordinates, and, A consist of the matrix of auxiliary fields. Parallel to equation (2.8) we may take the action integral in superspace as

$$S = \int dt\, d\theta\, d\bar{\theta}\, Tr\left[-\Phi\bar{D}D\Phi + W(\Phi)\right]. \tag{14.14}$$

$W(\Phi)$ above is the matrix superpotential, which is the negative of $V(F)$ of equation (2.8). We shall take this as

$$W(\Phi) = \frac{1}{2}\Phi^2 - \frac{\lambda}{3\sqrt{N}}\Phi^3. \tag{14.15}$$

We have retained a $\frac{1}{\sqrt{N}}$ in the above to consider the limit of N approaching infinity, which corresponds to the large N limit of field theory. The relevance of taking a matrix model as above is that the matrix model at critical points becomes equivalent to string theories as verified through explicit calculations. Although this understanding is yet incomplete, we shall discuss the model arising from the above as an example of supersymmetric quantum mechanics with many supercoordinates, which, leads to some interesting results.

In Chapter 2 we had eliminated the auxiliary coordinates and then written down the Hamailtonian. We shall do the same here. Then we obtain the action integral as

$$S = \int dt \left[Tr\left(-X\frac{d^2X}{dt^2} + F(X^2)\right) + \bar{\Psi}\frac{dF(X)}{dX}\Psi\right], \tag{14.16}$$

where

$$F(X) = X - \frac{\lambda}{\sqrt{N}}X^2. \tag{14.17}$$

Clearly, the above expression corresponds to the Hamiltonian in equation (2.26), with v' corresponding to $F(X)$ with a change of sign. From equation (2.26) for $N = 1$ we then have the Hamiltonian breaking up into two sectors, with the bosonic sector as

$$H^{(B)} = p^2 + (F(x))^2 - \frac{dF(x)}{dx}. \qquad (14.18a)$$

From the above clearly we have for the bosonic sector the potential $V(x)$ is given as

$$V(x) = [x(1 - \lambda x)]^2 - 1 + 2\lambda x. \qquad (14.18b)$$

The Hamiltonian for the fermonic sector with $N = 1$ is given as

$$H^{(F)} = p^2 + (F(X))^2 + \frac{dF(x)}{dx}. \qquad (14.19)$$

In the above the difference in normalisation for the kinetic term may be noted. The reason for omitting the factor $(1/2)$ in equations (14.16) and (14.18) is that the WKB approximation to be used later will look simpler. For any N, the Hamiltonian of the bosonic sector has a form parallel to the above, and is given as

$$H^{(B)} = Tr\left[P^2 + (F(X))^2 - \frac{dF(X)}{dX}\right]. \qquad (14.20)$$

A bosonic matrix model similar to the above was considered quite some time back [12] where the ground state wave function or the corresponding vacuum structure was obtained. We note that X as a hermitian matrix is equivalent to N^2 bosonic variables. When we consider the wave function Ψ for the vacuum, we would like it to maintain the symmetry of the Hamiltonian. Hence we should have

$$\Psi(X) = \Psi(U(N)XU^{-1}(N)). \qquad (14.21)$$

Hence the wave function Ψ will not be a function of the N^2 variables present in X, but will depend only on a subset of them. This is similar to a rotationally symmetric wave function being dependent only on $|\vec{x}|$ and not on the three variables in $\vec{x}$. In fact, if $(\lambda_1, \cdots, \lambda_N)$ are the eigenvalues of the matrix X, then the wave function Ψ as above will be a symmetric function of only the above N eigenvalues [12]. On evaluation of the expectation value of vacuum energy, it has been further shown that [12] the substitution

$$\phi(\lambda_1, \cdots, \lambda_N) = \left\{\prod_{i<j}(\lambda_i - \lambda_j)\right\}\psi(\lambda_1, \cdots, \lambda_N) \qquad (14.22)$$

14.3. Matrix Models

will yield the correct vacuum structure, where ϕ corresponds to a system of noninteracting fermions subjected to the common potential

$$V = N\,V(x/\sqrt{N}). \tag{14.23}$$

Here $V(x)$ is the same potential as given by equation (14.18b). We note that the wave function ϕ being antisymmetric in the arguments makes it fermionic. We therefore now examine the dynamical system of N fermion subjected to the potential (14.23).

The ground state energy ϵ for the matrix model is now given in terms of the single particle energies $e_1, \cdots, e_N$ for the fermions. Thus we have [12]

$$\mathcal{E} = \sum_{k=1}^{N} e_k. \tag{14.24a}$$

Let e_F be the Fermi surface in the space of fermions. We may then rewrite the above equation as

$$\mathcal{E} = \sum_k \theta(e_F - e_k) e_k. \tag{14.24b}$$

The particle number N is also given as

$$N = \sum_k \theta(e_F - e_k). \tag{14.25}$$

On using WKB approximation and with a rescaling of energy variables we then obtain from equation (14.24b) that [12]

$$\mathcal{E} = e_F - \frac{3}{2\pi} \int dx \left[e_F - V(x,\lambda)\right]^{3/2}. \tag{14.26}$$

In the above we have explicitly written the dependance of the potential on λ. Also in this approximation equation (14.25) becomes [12]

$$1 = \frac{1}{\pi} \int dx \left[e_F - V(x,\lambda)\right]^{1/2}. \tag{14.27}$$

In equations (14.26) and (14.27), the range of integration is over those values of x for which the integrand is real. The Fermi level for the free fermion system is clearly determined from equation (14.27). We shall now consider the case when the expression equation (14.26) shall have a singularity.

Let us now consider the behavior of the potential V(x) of equation (14.18b). It is clear that for small λ and the potential $V(x)$ has three extrema x_1, x_2, x_3 with x_1 and x_3 being minima, and, x_2 being a maximum. Also, when λ is large, we have a single minimum at x_1. At a critical value $\lambda = \lambda_c$, we obtain that $x_2 = x_3$. We can estimate this critical value as $\lambda_c = \frac{1}{6}\sqrt{3}$. We have two or one minimum depending on whether $\lambda < \lambda_c$ or $\lambda > \lambda_c$. One may verify numerically [11] that for small values of λ, the solution of equation 14.27) for e_F yields that

$$V(x_3) = e_F. \qquad (14.28)$$

This may be compared with the usual result of supersymmetric quantum mechanics when the potential well at x_1 will describe the eigenvalues corresponding to $H^{(B)}$, whereas the potential well with x_3 as a minimum describes the eigenvalues of $H^{(F)}$. The energy levels of fermionic and bosonic states match except for the lowest bosonic mode corresponding to 'vacuum'. In contrast to the above the present description of the matrix model is not manifestly supersymmetric, as may be particularly seen for $\lambda > \lambda_c$ [11]. Equation (14.26) has a singularity [12] when $\lambda = \lambda_c$, and then the single particle levels are indefinitely closely spaced near e_F.

It has been noted [11,13] that the matrix model with the double scaling limit of $N \to \infty$ and $\lambda \to \lambda_c$ describes the continuum behavior of string theory and two dimensional gravity, topics which are of primary interest at present as facets of probable future theories. At present this is the main reason for the interest in matrix models, although as yet the exact relationship between them is not fully understood. Here we have not dealt with the calculations from two dimensional surfaces corresponding to surfaces of genus n, or the summation of the contributions from these surfaces with different topologies. This calculation will need a more careful analysis of the double limit $N \to \infty$ and $\lambda \to \lambda_c$ [13] as well as the expectations from conformal field theory, which has been omitted here.

Further aspects of nonperturbative solutions of the discretised Green Schwarz superstring in one dimensions has also been considered in Ref.[14] as a direct continuation of the above. This is at present an active field of research in the context of matrix models as such as well as in the context of string theory and conformal field theory.

14.4 Remarks

We note that the consideration of gravity sector as it results from supergravity is relevant both from a conceptual as well as phenomenological point of view. In the context of inflationary universe [15] or the pre-Planckean era of the universe in the big bang cosmology, supergravity as a quantum theory should be relevant. But it is so complicated that one does not even contemplate including it yet. One deals with Einsteinian gravity in a classical manner in all such problems and hope that this may be adequate. However, for the sake of aesthetics as well as completeness, theoretical insight into such problems will be nice. For example, one deals with graviton scattering based on string theory [16], which for example gives a better insight for superstring based supergravity with a manifestly physical process. Further work in the same direction has been done by Gava, Iengo and Zhu [17]. Here it was shown that a two loop calculation for four bosonic massless fields with suitable compactification generates ultrviolet finite corrections for quantum gravity because of its origin from superstring theory. The theory however still has infrared divergences. Can it be resolved because the universe is finite in size? Such a calculation with Feynman diagrams can not be even suitably defined, and therefore identification of these results through conventional perturbation theory is not possible. There have been many attempts to use nonperturbative dynamics with string based supergravity. These show, albeit in hypothetical problems, the beauty of supersymmetry and supergravity being outputs instead of being inputs. We have mainly addressed ourselves to some simple aspects of these ideas in the last four Chapters. Further work in this direction is actively pursued. We have exhausted the limits of perturbative dynamics, and are now fully aware of its limitations. Our next stage of understanding will need the development of nonperturbative methods where the options are many. The application of these techniques to supersymmetry and supergravity is not simple, and, is only partly resolved. As a theory similar to quantum field theory giving quantitative results, we are as yet nowhere near a solution. Such a solution must be found.

REFERENCES

1. S. Coleman and J. Mandula, Phys. Rev. 159, 1251 (1967).

2. H. Haber and G. Kane, Phys. Rep. 117, 76 (1984).

3. T.G. Rizzo, Phenomenology of E_6 superstring inspired models, MAD/PH/526 (1989).

4. J.C. Pati, Phys. Lett. 228B, 228 (1989).

5. J.C. Pati, M. Cvetic, and H.S. Saratchandra, Phys. Rev. Lett. 58, 851 (1987).

6. R.N. Mohapatra and J.C. Pati, Phys. Rev. D11, 566 (1975).

7. M. Gell Mann, Phys. Rev. 125, 1067 (1962).

8. H. Georgi, H. Quinn and S. Weinberg, Phys. Rev. Lett. 33, 452 (1974).

9. S. Mahapatra and S.P. Misra, Phys. Rev. D33, 3464 (1986); S. Mishra and S.P. Misra, Phys. Lett. 186B, 99 (1987).

10. E. Witten, Nucl. Phys. B185, 513 (1981), K. Konishi, Phys. Lett. 135B, 439 (1984).

11. E. Marinari and G. Parisi, Phys. Lett. 240B, 375 (1990); G. Parisi, Phys. Lett. 238B, 209 (1990).

12. E. Brezin, C. Itzykson, G. Parisi and J.B. Zuber, Commun. Math. Phys. 59, 35 (1978).

13. D.J. Gross and N. Miljkovic, Phys. Lett. 238B, 217 (1990); G. Parisi, Phys. Lett. 238B, 213 (1990).

14. S. Bellucci, T.R. Govindrajan, A. Kumar and R.N. Oerter, Phys. Lett. 249, 49 (1990).

15. A. Guth, Phys. Rev. D23, 374 (1981); A Linde, Phys. Lett. 129B, 177 (1983).

16. D. Amati, M. Ciafaloni and G. Veneziano, Phys. Lett. B 197, 81 (1988); Int. J. Mod. Phys. A3, 1615(1988).

17. E. Gava, R. Iengo and C. Zhu, Nucl. Phys. B323, 585 (1989).

Appendix A

Construction of Calabi Yau Manifolds

We shall briefly consider here the construction of Calabi-Yau manifolds which are relevant in the context of compactification of the ten dimensional supergravity theories to four dimensions [1]. We shall start from the beginning [2-5] and discuss up to Complete Intersection Calabi Yau (CICY) spaces, which are the spaces mainly used for phenomenology of superstring based theories. Most of what we discuss here is contained in Ref [2,3].

A.1 Manifolds

We first note that a manifold is a topological space which is **locally Euclidean**. If the space has even number of dimensions, say $2n$, it can be described by n *complex* coordinates z^μ ($\mu = 1, \cdots, n$). We shall recognize the $2n$ degrees of freedom by formally considering $2n$ complex coordinates $\{z^\mu, z^{\bar\mu} : \mu = 1, \cdots, n\}$. The $2n$ dimensional space should be locally Euclidean. Hence at every point there must be a **neighborhood** U_i which is homeomorphic to a subspace of a Euclidean space, and thus points of U_i are described by a coordinate system z^μ borrowed from the points of the Euclidean space. If there is another neighborhood U_j with a coordinate system w^μ with the intersection between U_i and U_j as nonempty, then there will be points of M which can be described by either of the two coordinate systems. Such a dual description should be smooth. We ensure this by the requirement that w^μ is an analytic function of the z's, and that the inverse of this function exists. We then call M a **complex manifold**. The U_i's with associated coordinates are called **maps** (or **charts**) and a set of such maps covering whole of M is called an **atlas**.

A.2 Complex structures

Let us consider the tensor

$$J = idz^\mu \otimes \frac{\partial}{\partial z^\mu} - idz^{\bar{\mu}} \otimes \frac{\partial}{\partial z^{\bar{\mu}}}. \qquad (A.1)$$

We recognize that dz^μ and $dz^{\bar{\mu}}$ constitute a basis for covariant tensors of rank one with complex indices, and similarly that $\frac{\partial}{\partial z^\mu}$ and $\frac{\partial}{\partial z^{\bar{\mu}}}$ form a basis of the contravariant tensors of rank one with complex indices. Thus, for the tensor given by equation (A.1), in terms of components we have, with indices corresponding to the complex coordinates,

$$J_\mu{}^\nu = i\delta_\mu{}^\nu; \quad J_{\bar{\mu}}{}^{\bar{\nu}} = -i\delta_{\bar{\mu}}{}^{\bar{\nu}}. \qquad (A.2)$$

With the present coordinate system, the other components of the tensor are zero. This tensor is globally defined as a numerical tensor for a given coordinate system. Further, we have

$$J_\mu{}^\nu J_\nu{}^\lambda = -\delta_\mu{}^\lambda, \quad J_{\bar{\mu}}{}^{\bar{\nu}} J_{\bar{\nu}}{}^{\bar{\lambda}} = -\delta_{\bar{\mu}}{}^{\bar{\lambda}}. \qquad (A.3a)$$

Let us now have a real coordinate system x^m with $m = 1, \cdots, 2n$. We then define an **almost complex structure** on M as a tensor $J_m{}^n$ which is globally defined on M such that

$$J_m{}^n J_n{}^p = -\delta_m{}^p, \qquad (A.3b)$$

a property which is satisfied by the tensor above. We note that as a matrix the square of the complex structure is negative of identity. An almost complex structure is said to be a **complex structure** if in addition to (A.3b), the **Niejenhuis tensor**

$$N_{ij}{}^k = J_{[i}{}^k{}_{,j]} - J_{[i}{}^p J_{j]}{}^q J_p{}^k{}_{,q} \qquad (A.3c)$$

vanishes. We can then show that there is always an atlas so that (A.2) is true.

A.3 Hermitian Complex Manifolds

On a manifold we can introduce a metric. On the complex manifold, let us introduce the **metric** given as

$$ds^2 = g_{\mu\bar{\nu}} dz^\mu dz^{\bar{\nu}}, \qquad (A.4)$$

with $g^*_{\mu\bar{\nu}} = g_{\nu\bar{\mu}}$. Then the complex manifold is **hermitian**. We note that the covariant metric tensor $g_{\mu\bar{\nu}}$ will retain this mixed form in complex indices under a coordinate transformation; i.e. components like $g_{\mu\nu}$ or $g_{\bar{\mu}\bar{\nu}}$ will always remain zero. With arbitrary coordinates the hermitian metric satisfies the relationship

A.4. Differential Forms

$$g_{mn} = J_m{}^k J_n{}^l g_{kl}. \qquad (\text{A.5})$$

Further, $J_{mn} = J_m{}^k g_{kn}$ is a covariant antisymmetric tensor of rank two which, as we shall just see, naturally defines a two form.

A.4 Differential Forms

On a complex manifold differential forms are defined as in Chapter 8. However here we shall for any form separately consider the dependance on dz^μ and $dz^{\bar\mu}$ and also explicitly use $\wedge$ for products of forms so as to distnguish the same from tensor products. Thus $\omega^{(p,q)}$ is a (p,q) form when there are p holonomic differentials and q antiholonomic differentials corresponding to the coordinates z^μ and $z^{\bar\mu}$ respectively. In fact, for the almost complex structure $J_m{}^n$ in a hermitian manifold we have as stated the natural $(1,1)$ form

$$\mathcal{J} = J_{\mu\bar\nu} dz^\mu \wedge dz^{\bar\nu}, \qquad (\text{A.6})$$

since $J_{\mu\bar\nu}$ is an antisymmetric tensor. We may recognise that J in (A.1) defines a mixed tensor, whereas $\mathcal{J}$ in (A.6) is a "mixed" two form.

We can now define two types of exterior derivatives ∂ and $\bar\partial$. Clearly $\partial \omega^{(p,q)} = \omega^{(p+1,q)}$, and $\bar\partial \omega^{(p,q)} = \omega^{(p,q+1)}$. The usual exterior derivative of the manifold is given as

$$d = \partial + \bar\partial. \qquad (\text{A.7})$$

Let $Z^{(p,q)}$ be all **closed** (p,q) forms, i.e.

$$Z^{(p,q)} = \{\omega^{(p,q)} : d\omega^{(p,q)} = 0\}, \qquad (\text{A.8a})$$

and, $B^{(p,q)}$ be all **exact** (p,q) forms, i.e.

$$B^{(p,q)} = \{\omega^{(p,q)} : \omega^{(p,q)} = d\nu \text{ for some } \nu\}. \qquad (\text{A.8b})$$

We define

$$H^{(p,q)} = Z^{(p,q)} / B^{(p,q)} \qquad (\text{A.8c})$$

as the corresponding factor group. Then,

$$dim\left(H^{(p,q)}\right) = b_{p,q} \qquad (\text{A.9})$$

are the **Betti-Hodge numbers** for the manifold M. During compactification, these numbers will give us the number of families as well as the number of antifamilies. The **Betti numbers** b_r are given as

$$b_r = \sum_{p+q=r} b_{p,q}. \qquad (\text{A.10})$$

A.5 Connection and Curvature

For the hermitian manifold defined in Section A.3, connection can be taken as

$$\Gamma_{\mu\nu}{}^{\kappa} = g^{\kappa\bar{\rho}}\partial_{\mu}g_{\nu\bar{\rho}}\,. \qquad (\text{A}.11)$$

We can then show that the covariant derivative of the metric vanishes, and that the covariant derivative of the complex structure also vanishes. In fact, either of these conditions determine the above connection uniquely for the hermitian manifold M. We may note that the connection is pure in its indices. With the curvature tensor generally given as

$$R_{mnl}{}^{k} = \partial_{[m}\Gamma_{n]l}{}^{k} + \Gamma_{[mr}{}^{k}\Gamma_{n]l}{}^{r}, \qquad (\text{A}.12)$$

we obtain that the only nontrivial components of curvature are given in terms of complex indices as

$$R_{\mu\bar{\nu}\bar{\rho}}{}^{\bar{\sigma}} = -R_{\bar{\nu}\mu\bar{\rho}}{}^{\bar{\sigma}} = \partial_{\mu}\Gamma_{\bar{\nu}\bar{\rho}}{}^{\bar{\sigma}}. \qquad (\text{A}.13)$$

The above expression enables us to define the **Ricci two form** given as

$$\mathcal{R} = iR_{\mu\bar{\nu}\bar{\rho}}{}^{\bar{\rho}}dz^{\mu} \wedge dz^{\bar{\nu}} \qquad (\text{A}.14a)$$

$$= i\partial\bar{\partial}\,log(g^{1/2}). \qquad (\text{A}.14b)$$

The metric and thus the determinant of the metric g are defined in each coordinate patch, but *the Ricci form can be shown to be defined globally*.

A.6 Kahler Manifold

A hermitian manifold is **Kahler** if $d\mathcal{J} = 0$, i.e. the two form defined by the complex structure as in equation (A.6) is closed. From this we can obtain that $\partial_{\mu}g_{\nu\bar{\rho}} = \partial_{\nu}g_{\mu\bar{\rho}}$ and that $\partial_{\bar{\rho}}g_{\mu\bar{\nu}} = \partial_{\bar{\nu}}g_{\mu\bar{\rho}}$ which are really integrability conditions for the metric. Hence, on a coordinate patch U_i we can define a "potential" ϕ_i called the **Kahler potential** such that

$$g_{\mu\bar{\nu}} = \partial_{\mu}\partial_{\bar{\nu}}\,\phi_i. \qquad (\text{A}.15)$$

Although $\mathcal{J}$ is closed, it is not exact. To see this let us simplify $\mathcal{J} \wedge \mathcal{J} \wedge \cdots \wedge \mathcal{J}$ with n factors. A little algebra then yields that

$$\mathcal{J} \wedge \mathcal{J} \wedge \cdots \wedge \mathcal{J}(n\text{ factors})$$
$$= i^{n}(n!)\,g\,dz^{1} \wedge dz^{\bar{1}} \wedge \cdots \wedge dz^{n} \wedge dz^{\bar{n}}. \qquad (\text{A}.16)$$

The right hand side clearly is proportional to the volume element of the manifold. If $\mathcal{J}$ were exact, then use of Stokes theorem would immediately give the volume of the manifold to be zero, which shall be unphysical and is not true.

A.7 Calabi-Yau Manifold

A Calabi Yau manifold is a compact Kahler manifold which is Ricci flat. Hence the Ricci tensor $R_{\mu\bar{\nu}}$ vanishes. Therefore in equation (A.14) the Ricci two form $\mathcal{R}$ vanishes. However, $\mathcal{R}$ is the first of a series of analytic invariants on the manifold constructed from the curvature two-form known as Chern classes. $\mathcal{R}/(2\pi)$ is in fact the first Chern class, and thus we may also define a Calabi Yau manifold as a compact Kahler manifold with vanishing first Chern class.

The most important property of the Calabi Yau manifold directly relevant to us is given by the following theorem:

Theorem: *A Calabi Yau manifold admits a globally defined one form $A = A_m dx^m$ and a pair of globally defined spinors $\zeta, \bar{\zeta}$ of opposite chirality such that*

$$(\nabla_m - \frac{i}{2} A_m)\zeta = 0, \qquad (A.17a)$$

$$(\nabla_m + \frac{i}{2} A_m)\bar{\zeta} = 0. \qquad (A.17b)$$

In the above, the one form A is a potential for the Ricci form so that we have $\mathcal{R} = dA$ and $dA = 0$ is the integrability condition for the global existence of A.

Let us now *confine our discussions to a six dimensional manifold* with three complex coordinates. This is relevant for us for compactification of the ten dimensional world of supergravity to the four observed dimensions, with six dimensions compactified to small scales so that this space is not accessible at present energies. Our description here is for the above six dimensional compact space, and, ∇_m in equations (A.17) includes the spin connection of this space. We explicitly have

$$\nabla_m \eta = \partial_m \eta - \frac{1}{4} \omega_m{}^a{}_b \gamma_a{}^b \eta, \qquad (A.17c)$$

where $\omega_m{}^a{}_b$ are the components of the connection one form and $\gamma^{ab} = [\gamma^a, \gamma^b]$.

Chirality of the fermions in this space is given through the matrix

$$\gamma = \frac{i}{6!}\epsilon_{mnpqrs}\gamma^m\gamma^n\gamma^p\gamma^q\gamma^r\gamma^s. \qquad (A.18)$$

With normalisation as $\zeta^\dagger\zeta = 1$, it can be shown that the complex structure here is given as

$$J_m{}^n = -i\zeta^\dagger\gamma_m{}^n\zeta. \qquad (A.19)$$

In the above equation we have reversed the roles; we have written an almost complex structure when a covariantly constant spinor is available. We may also write the γ-matrices with *complex* indices. With the metric as in equation (A.4) we then have, with $\mu, \nu = 1, 2$ or 3,

$$[\gamma^\mu, \gamma^\nu]_+ = 0 = [\gamma^{\bar\mu}, \gamma^{\bar\nu}]_+ , \qquad (A.20a)$$

and

$$[\gamma^\mu, \gamma^{\bar\nu}]_+ = 2g^{\mu\bar\nu}. \qquad (A.20b)$$

The above equations tell us that we can use the γ^μ-matrices as creation and annihilation operators. In fact we have from equations (A.19) and (A.5)

$$ig_{\mu\bar\nu} = -\zeta^\dagger\gamma^{\mu\bar\nu}\zeta$$

$$= -i\zeta^\dagger\left(\gamma^\mu\gamma^{\bar\nu} - g_{\mu\bar\nu}\right)\zeta,$$

which with a little algebra yields that

$$\gamma_{\bar\mu}\zeta = 0, \quad \gamma^\mu\zeta = 0. \qquad (A.21)$$

Thus ζ is the state of highest weight. We can construct the other states by using γ_μ or $\gamma^{\bar\mu}$ as the creation operators. We can then write an arbitrary spinor as

$$\begin{aligned}\eta &= \omega^{(0,0)}\zeta + \omega^{(0,1)}{}_{\bar\mu}\gamma^{\bar\mu}\zeta + \omega^{(0,2)}{}_{\bar\mu\bar\nu}\gamma^{\bar\mu\bar\nu}\zeta \\ &\quad + \omega^{(0,3)}{}_{\bar\mu\bar\nu\bar\rho}\gamma^{\bar\mu\bar\nu\bar\rho}\zeta.\end{aligned} \qquad (A.22)$$

From the above it is clear that under rotations of $SO(6) \cong SU(4)$, the eight components of the spinor break up under $SU(3)$ subgroup of $SU(4)$ into a singlet, a triplet, another triplet and a singlet, i.e. $8 = 4+\bar{4} = (1+3) + (\bar{3}+1)$. This reflects that the Calabi Yau space has $SU(3)$ holonomy. For vectors, we merely have $6 = 3 + \bar{3}$. We thus have e.g.

$$A = A_\mu dz^\mu + A_{\bar\mu} dz^{\bar\mu} \qquad (A.23)$$

and thus $\mathcal{R} = \partial(A_\mu dz^\mu) + \bar\partial(A_{\bar\mu} dz^{\bar\mu})$ as a result of $\mathcal{R} = dA$.

A.8 Betti Hodge numbers

As already mentioned in equation (A.9) the Betti-Hodge numbers b_{pq} of a complex manifold are given as the dimension of the space $Z^{(p,q)}/B^{(p,q)}$ where $Z^{(p,q)}$ and $B^{(p,q)}$ are respectively the spaces of closed and exact (p,q) forms. For a complex manifold it is useful to write these in an array called the **Hodge diamond** as below which reflects some symmetry properties:

$$\begin{array}{ccccccc}
 & & & b_{00} & & & \\
 & & b_{10} & & b_{01} & & \\
 & b_{20} & & b_{11} & & b_{02} & \\
b_{30} & & b_{21} & & b_{12} & & b_{03} \\
 & b_{31} & & b_{22} & & b_{13} & \\
 & & b_{32} & & b_{23} & & \\
 & & & b_{33} & & &
\end{array}$$

We have in the above $b_{pq} = b_{qp} = b_{n-q,n-p}$ and, $b_{p0} = b_{n-p,0}$. For the Calabi Yau manifold we have $b_{00} = 1$ and $b_{10} = b_{20} = 0$. Hence the only numbers which depend on the specific construction of the manifold here are b_{11} and b_{21}.

A.9 Examples of Calabi-Yau space

It is useful to construct Calabi-Yau spaces with polynomial intersections. Let us consider the space CP^n. This is the space with $n+1$ homogeneous complex coordinates z^A, $(A = 1, \cdots, n, n+1)$ with the identification that for any complex λ, $z \equiv \lambda z$. This projective space has dimension $2n$. Let us have N homogeneous polynomials $P_1, P_2, \cdots, P_N$ respectively of degrees $d_1, d_2, \cdots, d_N$. With $n = N + 3$, it clear that the subspace of CP^n obtained by the intersection of all the above polynomials equated to zero has three complex dimensions. It is a manifold if it is a complete intersection space, i.e. if

$$\mathcal{N} = dP_1 \wedge dP_2 \wedge \cdots \wedge dP_N \qquad (\text{A.24})$$

never vanishes at any point of the manifold. We can understand the above statement when we recognize that N is the normal to the hypersurface of the intersections, and having a nonzero normal ensures the smoothness of the surface. We shall construct Calabi Yau spaces from such complete intersection spaces immersed in CP^n. For this purpose we need that the first Chern class vanishes. We shall examine the condition which will ensure this. We shall call such spaces as **Complete Intersection Calabi Yau spaces** (CICY).

In the projective space, let us consider the $N+3$ form μ formally given as

$$\mu = \epsilon_{A_1 A_2 \cdots A_{N+4}} \, dz^{A_1} \wedge dz^{A_2} \wedge \cdots \wedge dz^{A_{N+4}}. \qquad (\text{A.25})$$

Under a scaling $z^A \to \lambda z^A$, we have

$$\mu \to \lambda^{N+4} \mu, \qquad (\text{A.26})$$

which implies that μ is not really defined in the projective space. We now consider the $N+3$ form defined as

$$\nu = \mu/(P_1 P_2 \cdots P_N). \qquad (\text{A.27})$$

We then construct the product of N contours corresponding to the above polynomials given as

$$\Gamma_N = \gamma_1 \otimes \gamma_2 \otimes \cdots \otimes \gamma_N,$$

and take the three form Ω given as

$$\Omega = \frac{1}{(2\pi i)^N} \int_{\Gamma_N} (\mu/P_1 P_2 \cdots P_N). \qquad (\text{A.28})$$

Since the degrees of the homogeneous polynomials are respectively $d_1, d_2, \cdots d_N$, under $z^A \to \lambda z^A$,

$$P_\alpha \to \lambda^{d_\alpha} P_\alpha.$$

For the complete intersection space, we now impose the additional condition that

$$\sum d_\alpha = N + 4. \qquad (\text{A.29})$$

This makes the form ν in the projective space as in equation (A.27) well defined, and thus the three form Ω in equation (A.28) also becomes well defined. To show that this defines a Calabi Yau space, we now use the following theorem:

Theorem: *A compact Kahler manifold has vanishing first Chern class if and only if the manifold admits a nowhere vanishing holomorphic three form.*

We have constructed in the above such a three form. Hence The first Chern class vanishes and thus the condition (A.29) makes the complete intersection space a Calabi Yau space.

The above construct does not have too many options. Firstly, we need not consider polynomials of degree one, since then we can eliminate one of the complex coordinates and effectively it amounts to considering CP^{n-1} space. Hence $d_\alpha \geq 2$. This gives rise to the restriction that

A.10. Examples of CICY spaces   215

$$N + 4 = \sum d_\alpha \geq 2N,$$

which in turn implies that

$$N \leq 4. \tag{A.30}$$

Hence we may consider [1] explicitly the following examples. (i) CP^4 with one polynomial of degree five, with Euler number as -200. (ii) CP^5 with two polynomials each of degree three, with Euler number -144, or one of degree two and the other of degree four with Euler number -176. (iii) CP^6 with three polynomials of degrees two, two and three and Euler number -144. (iv) CP^7 with four polynomials, each of degree two with Euler number -128. This exhausts all the available choices. We shall later give an expression for the calculation of Euler numbers when the above numbers as quoted may be verified.

A.10 Examples of CICY spaces

In order to illustrate the above technique in the context of applications we shall consider here two familiar examples, one given in Ref [1] and the other considered by Tian and Yau [6]. These are considered in Chapter 12 to describe compactification of ten dimensional supergravity, and also include additional discrete symmetries in constructing the space. The first one gives a four generation model, whereas, the second one gives a three generation model for quarks and leptons.

A.10.1 CHSW Manifold

Let us consider CP^4 space with the simple quintic polynomial equation [1]

$$\sum_{i=1}^{5} z_i^5 = 0 \tag{A.31}$$

Let the complete intersection Calabi Yau manifold thus obtained be Y_0. We now further introduce a discrete symmetry $Z_5 \times Z_5$ in this space generated by the elements A and B acting in a cyclic manner as

$$A : (z_1, z_2, z_3, z_4, z_5) \to (z_5, z_1, z_2, z_3, z_4), \tag{A.32a}$$

$$B : (z_1, z_2, z_3, z_4, z_5) \to (\alpha z_1, \alpha^2 z_2, \alpha^3 z_3, \alpha^5 z_4, z_5). \tag{A.32b}$$

In the above, α is a nontrivial fifth root of unity. Clearly A and B *donot* generate fixed points, and thus the discrete group acts freely on the manifold. We then consider the Calabi Yau space $Y = Y_0/(Z_5 \times Z_5)$. Because of the discrete symmetry, loosely speaking twentyfive pieces of the original space become a single piece of Y. The Betti-Hodge numbers may be calculated as $b_{11} = 1$ and $b_{21} = 5$. We then obtain, with an expansion parallel to equation (A.22), that there are one set of spinors of one chirality, and, five sets of spinors of opposite chirality. Thus the number of generations which equals the magnitude of the number of families minus the number of antifamilies, becomes four. In fact, we have, for the Calabi Yau manifold

$$\text{No. of families} = (1/2)\text{Euler no. of the manifold}$$

$$= |b_{11} - b_{21}|, \qquad (A.33)$$

a result which we are stating here without proof. Qualitatively it is based on equations like (A.22) giving an expansion of the Dirac spinor in the internal space with opposite chiralities as stated there. This manifold was considered by Candelas, Horowitz, Strominger and Witten (CHSW) as the first example for superstring based grand unification.

A.10.2 Tian Yau Manifold

We shall now consider the Tian Yau manifold [6], which has the following features. Here (i) we shall start with the product space $CP^3 \times CP^3$ and (ii) take a built in discrete symmetry Z_3 as below. The projective space here contains eight homogeneous coordinates, and thus as in the last subsection, we shall choose one quadratic and two cubic polynomial conditions for the complete intersection space to be a Calabi Yau space. The property (ii) as above is always essential for the consideration of models relevant to phenomenology as we have just seen in the previous model so that, the number of families is limited to either three or four.

We take the complex coordinates as x^A, y^A ($A = 1, \cdots, 4$), and the polynomials for describing the complete intersection as

$$\sum_{A=1}^{4} x^A y^A = 0, \qquad (A.34a)$$

$$\sum_{A=1}^{4} (x^A)^3 = 0, \qquad (A.34b)$$

A.11. Remarks

$$\sum_{A=1}^{4}(y^A)^3 = 0. \qquad (\text{A.34c})$$

We introduce a discrete symmetry similar to the last model where we take it to be generated by

$$(x^1, x^2, x^3, x^4) \otimes (y^1, y^2, y^3, y^4) \to$$

$$(x^2, x^3, x^1, \omega x^4) \otimes (y^2, y^3, y^1, \omega^2 y^4). \qquad (\text{A.35})$$

This Calabi Yau manifold has $b_{11} = 6$ and $b_{21} = 9$, as discussed later with some modifications in subsection A.11.3, and in Chapter 12 leading to three generations with Euler number -6.

A.11 Remarks

We have very briefly given the way in which complete intersection Calabi Yau manifolds are constructed. Most of the applications deal with this technique for the compactification of the ten dimensional supergravity theories to the phenomena in four dimensions as we may observe it as discussed in Chapter 12. We shall presently end this appendix with a few more definitions which are utilised quite often, as well as calculate b_{21} by considering deformation of complex manifolds.

A.11.1 Harmonic Forms

When a manifold admits a metric, we can define Hodge $*$ operation by

$$*(dx^{m_1} \wedge dx^{m_2} \wedge \cdots \wedge dx^{m_p})$$

$$= \frac{1}{(n-p)!} g^{1/2} g^{m_1 k_1} \cdots g^{m_p k_p} \epsilon_{k_1 \cdots k_p k_{p+1} \cdots k_n} dx^{k_{p+1}} \wedge \cdots \wedge dx^{k_n}. \qquad (\text{A.36})$$

This enables us to define a scalar product of two p-forms given as

$$(\alpha_p, \beta_p) = \int \alpha_p \wedge *\beta_p$$

$$= \frac{1}{p!} \alpha_{m_1 \cdots m_p} \beta^{m_1 \cdots m_p} g^{1/2} dx^1 \wedge \cdots \wedge dx^n. \qquad (\text{A.37})$$

With respect to the above definition of the scalar product, we can define the hermitian conjugate of the operator d as $d^\dagger$ explicitly given as

$$d^\dagger = (-1)^{p(n-p+1)} * d *.\qquad(A.38)$$

We note that $d^\dagger$ takes a p-form to a $p-1$ form. A Laplacian operator Δ is then defined as

$$\Delta = dd^\dagger + d^\dagger d.\qquad(A.39)$$

A differential form ω is said to be harmonic if

$$\Delta\omega = 0.\qquad(A.40)$$

It can be shown that any differential form ω on a compact manifold without boundary can be written as

$$\omega = \alpha + d\beta + d^\dagger\gamma,\qquad(A.41)$$

where the first form is harmonic, the second is exact, and the third is co-exact. The above is known as **Hodge decomposition theorem**.

A.11.2 Chern Classes and Euler Characteristic

Chern classes are analytic invariants constructed from the Kahler form $\mathcal{J}$. For the complete intersection manifold as in subsection A.9, let us consider the expression

$$c = \frac{(1+\mathcal{J})^{N+4}}{\prod_\alpha (1 + d_\alpha \mathcal{J})}.\qquad(A.42)$$

The above expression is to be understood as a polynomial in $\mathcal{J}$ where we note that $\mathcal{J}^k = 0$ when $k > 3$. We can then write equation (A.42) as

$$c = 1 + c_1 + c_2 + c_3,\qquad(A.43)$$

with c_k as a $2k$ form containing $\mathcal{J}^k$. c_k is the kth **Chern class**. It may be recognised that we have not defined the Chern class as such, but in the above have stated what it is for the specific CICY space. The general definition depends on the fact that they are analytic invariants which characterise some topological properties of the manifold. We may note that c_1 for CICY spaces is zero since condition (A.29) is satisfied. This was in fact a prerequisite for the manifold to be a Calabi Yau space. **Euler number** $\chi(M)$ of the manifold M is defined here in terms of the third Chern class and is given as

$$\chi(M) = \int c_3.\qquad(A.44)$$

A.11. Remarks

Thus, from equation (A.42), the Euler number of the manifold is given by the coefficient of $\mathcal{J}^3$ in the above expansion, with correction factors from the intersecting polynomials. A direct evaluation shows that from equation (A.42),

$$c_3 = \frac{1}{3}\Big[(N+4) - \sum_\alpha d_\alpha{}^3\Big]\mathcal{J}^3. \qquad (\text{A.45})$$

For the extra factors coming from the polynomials for the purpose of integration of the above expression, we note that each polynomial condition of homogeneous degree d_α gives rise to d_α Rieman sheets. Taking into account this, from equations (A.44) and (A.45) we obtain that for the complete intersection Calabi Yau manifolds constructed from CP^{N+3},

$$\chi(M) = \frac{1}{3}\Big[(N+4) - \sum_\alpha d_\alpha{}^3\Big]\prod(d_\alpha). \qquad (\text{A.46})$$

Using the above formula, we can verify the Euler numbers for the manifolds as quoted at the end of (A.30). The Euler number is very significant for us since the number of generations equals $|b_{11} - b_{21}|$ which in turn is $\frac{1}{2}|\chi(M)|$.

We also have freely acting discrete symmetry groups G operative for the manifolds as in the examples of section A.10. This decreases the corresponding numbers by the factor which equals the number of elements of the discrete groups. Thus, if $n(G)$ is the order of the group G, then the Euler number of the space M/G is given as

$$\chi(M/G) = \chi(M)/n(G). \qquad (\text{A.47})$$

The product of the degrees of the polynomials increases the Euler number making it very large. We attempt to compensate for this by taking multiply connected space M/G using a freely acting group G so that we may have a three or four generation model for quarks and leptons as may be admissible.

The Euler number for the complete intersection Calabi Yau spaces constructed by starting from *products* of CP^n spaces can be considered in a similar way through the corresponding Chern class c_3. For the Tian Yau manifold, which also belongs to this category we note that $\chi = -6$ which leads to three generations of quarks and leptons. As stated earlier, the constructions of quarks and lepton generations will depend on the two Betti-Hodge numbers b_{11} and b_{21}. We next proceed to see how these can be calculated.

A.11.3 Betti-Hodge numbers and Complex Deformations

The calculations of the Betti-Hodge numbers depends on deformation theory [4] and some other technical results [3-5]. We shall here give only an introduction to the same [3]. We shall confine our discussions to the Tian Yau manifold [6] only. We first quote the result [6] that if we have V constructed from $CP^3 \times CP^3$ with polynomial conditions of degrees $(n,0)$, $(0,m)$ and $(1,1)$, then the Euler number is given as [6]

$$\chi(V) = -2nm(4-n)(4-m). \qquad (A.48)$$

Thus, as in section A.10, with $n = m = 3$, the Euler number for the manifold K_0 becomes -18.

Let us note that the Betti-Hodge numbers for $M = CP^n$ manifolds are given as

$$b_{pq} = \begin{cases} \delta_{p,q} & p \leq n \\ 0 & p > n \end{cases} \qquad (A.49)$$

Also, for the same for product manifolds $M \times N$, Kunneth formula given as

$$b_{pq}(M \times N) = \sum_{rmns} b_{rs}(M) . b_{mn}(N), \qquad (A.50)$$

subject to the constraints $r + m = p$ and $s + n = q$, shall be useful. We can obtain the Betti-Hodge numbers for the Tian Yau manifold from the above by applying some other formulae of Lefshetz. However, as already stated, we shall proceed with a consideration of deformations which gives us a more powerful method for getting them.

If the definition of a family of manifold of dimension n contains k complex parameters, then we get a complex analytic family of manifold of dimension $n + k$ including the space of parameters when certain mathematical properties are satisfied. A given manifold with specific values of the parameters shall be the submanifold of the larger manifold. Under deformations, we shall naturally get other members of the family. An interesting property which is very useful to us, is that, the effective number of parameters under deformation (or the dimension of the moduli space for the family of manifolds) equals b_{21}. We shall employ this method for the determination of b_{21}. Since as stated in section A.8, b_{21} and b_{11} are the only quantities to be determined for a Calabi Yau manifold, if we have a way of determination of the Euler number as in equation (A.48), b_{11} is automatically also known.

Let us consider the manifold CP^3 constrained by a single cubic polynomial $F = a_{ABC} x^A x^B x^C$. The number of parameters a_{ABC} is

A.11. Remarks

$^4C_1 + 2\ ^4C_2 + ^4C_3 - 1 = 19$ where we recognise that the space is a projective space so that one parameter as taken above is redundant and thus is to be subtracted. However, the dimension of the general linear transformations in the projective space is $16 - 1 = 15$. Hence the effective number of complex parameters for the definition of the manifold is $19 - 15 = 4$. Similarly, the second set of the homogeneous coordinates with the cubic constraint for the Tian Yau manifold of subsection A.10.2 gives rise to four effective complex parameters. We may thus take the two cubic polynomials corresponding to the Tian Yau manifold with only eight complex parameters as

$$\sum_A (x^A)^3 + a_1 x^2 x^3 x^4 + a_2 x^3 x^4 x^1 + a_3 x^4 x^1 x^2 + a_4 x^1 x^2 x^3 = 0,$$

(A.51a)

$$\sum_A (y^A)^3 + b_1 y^2 y^3 y^4 + b_2 y^3 y^4 y^1 + b_3 y^4 y^1 y^2 + b_4 y^1 y^2 y^3 = 0,$$

(A.51b)

For the quadratic constraint involving homogeneous coordinates for both the CP^3, use of the transformations of the homogeneous coordinates is no longer available. We might thus take it generally as

$$\sum_{A,B} c_{AB} x^A y^B = 0,$$

(A.52)

where, since it is a projective space, we take $c_{00} = 1$, thus leaving 15 effective complex parameters in the definition of this polynomial. The space K_0 for the Tian Yau manifold thus has 8+15=23 deformation parameters, which is the dimension of the moduli space and thus is b_{21} as stated above. The 23 monomials in x and y which multiply the 23 complex parameters a_A, b_A, c_{AB} (excluding c_{00} which we substitute without loss of generality as 1) in equations (A.51) and (A.52) correspond to a basis of the space $H^{(2,1)}$ as in equation (A.9) of section A.4. This completes the consideration of b_{21} for K_0 of the Tian Yau manifold through deformations. We now proceed to consider the same for $K = K_0/G$.

We take the discrete group $G = Z_3$ acting on K_0 to be generated through the transformation of the homogeneous coordinates given as [3]

$$g(x^1, x^2, x^3, x^4; y^1, y^2, y^3, y^4)$$
$$= (x^1, \alpha^2 x^2, \alpha x^3, \alpha x^4; y^1, \alpha y^2, \alpha^2 y^3, \alpha^2 y^4),$$

(A.53)

where $\alpha = exp(2\pi i/3)$ is a cube root of unity. Now, among the 23 monomials of K_0 stated in the last paragraph, the following nine monomials

$$\lambda_1 = x^1 x^2 x^3, \; \lambda_2 = x^1 x^2 x^4, \; \lambda_3 = y^1 y^2 y^3, \; \lambda_4 = y^1 y^2 y^4,$$

$$\lambda_5 = x^2 y^2, \; \lambda_6 = x^3 y^3, \; \lambda_7 = x^4 y^4, \; \lambda_8 = x^3 y^4, \; \lambda_9 = x^4 y^3, \quad (\text{A.54a})$$

are invariant under Z_3. Thus they form a basis for $H^{(2,1)}$ of $K = K_0/Z_3$. Among the remaining monomials, the seven monomials

$$Q_1 = x^2 x^3 x^4, \; Q_2 = y^1 y^3 y^4, \; Q_3 = x^1 y^2, \; Q_4 = x^2 y^3,$$

$$Q_5 = x^2 y^4, \; Q_6 = x^3 y^1, \; Q_7 = x^4 y^1, \quad (\text{A.54b})$$

get multiplied by α, and the seven monomials

$$Q_1^c = x^1 x^3 x^4, \; Q_2^c = y^2 y^3 y^4, \; Q_3^c = x^1 y^3, \; Q_4^c = x^1 y^4,$$

$$Q_5^c = x^2 y^1, \; Q_6^c = x^3 y^2, \; Q_7^c = x^4 y^2, \quad (\text{A.54c})$$

get multiplied by α^2. Through the mechanism of symmetry breaking through flux quantisation or Wilson loops, the monomials in equations (A.54a), (A.54b) and (A.54c) in K will have residual group symmetry as in equation (12.46) of Chapter 12. Hence, the effective number of deformation parameters in K is nine, and thus $b_{21}(K) = 9$. We can calculate b_{11} also by using Lefschetz's fixed point theorem [3,5]. Instead, we shall use the expression for the Euler number $\chi(K) = -6 = -2(b_{21} - b_{11})$ as from equations (A.47) and (A.48) so that from relations already stated we obtain that $b_{11} = 6$.

Parallel to equations (A.54), we should also consider a basis for $H^{(1,1)}$, which will on compactification describe the particles of same symmetry structure but opposite chirality in a manner similar to the expansion in equation (A.22) for spinors, and thus will give modes corresponding to "antigenerations". We discuss this in Chapter 12 when we associate the degrees of freedom of equations (A.54) with quarks and leptons so that the particles corresponding to $H^{(1,1)}$ will correspond to "mirror" quarks and leptons, the difference between the original and the mirror particles here being the number of generations.

Without any explanation, we give here a general statement related to the above [4]:

Let M be a compact complex manifold. If the number of moduli $m(M)$ is defined, then $m(M) = dim H^1(M, \Theta)(= b_{21})$.

Loosely speaking, the moduli above are the effective deformation parameters for the complex manifold.

A.11. Remarks

In the above we have considered the topological structure of the manifold. The explicit geometry of the manifold will depend on the values of the effective parameters which multiply the monomials in equations (A.51) and (A.52). These will in turn influence some physical quantities like Yukawa coupling [7], which will involve integration over the manifold K. Integrations over the manifolds will also give rise to calculable Yukawa couplings. We shall discuss some of this in the context of physically viable models in section 4 of Chapter 12 as well as the physical significance of purely internal degrees of freedom of compact space as in equations (A.54).

REFERENCES

1. P Candelas, G. Horowitz, A Strominger and E. Witten, Nucl. Phys. B258, 46 (1985).

2. P. Candelas, *Superstrings '87, Proceedings Trieste Spring School*, edited by Alvarez-Gaume et al, World Scientific, Singapore, 1987, pg.1; T. Eguchi, P.B. Gilkey and A.J. Hanson, Phys. Rep. 66C, 213 (1980).

3. B.R. Greene, K.H. Kelley, P.J. Miron, G.G. Ross, Nucl. Phys. B278, 667 (1986); B292, 606 (1987); P. Candelas, Nucl. Phys. B298, 458 (1988).

4. K. Kodaira, *Complex Manifolds and Deformation of Complex Structures*, Springer-Verlag, New York (1986).

5. P. Griffiths and J. Harris, *Principles of Algebraic Geoemtry*, Wiley, 1978; R.C. Gunning and H. Rossi, *Analytic Foundations of Several Complex Variables*, Prentice Hall, Englewood Cliffs N.J., 1965; K. Yano, *Differential Geometry on Complex and Almost Complex Spaces*, MacMillan, New York, 1965.

6. G. Tian and S.T. Yau, *Proceedings of the Argonne Symposium on Anomalies, Geometry and Topology*, edited by W.A. Bardeen and A.R. White, World Scientific, Singapore, 1985, p.402.

7. S. Kalara and R.N. Mohapatra, Phys. Rev. D 35, 3145 (1987); D36, 3474 (1987).

Appendix B

Some Results in Group Theory

We shall consider here some results in group theory which are needed for describing compactification of the ten dimensional space of supergravity to four dimensions and for grand unification theories. We shall briefly quote only those results without which a discussion of the symmetry structure in four dimensions shall not be meaningful. We shall exclusively use the materials of Ref. [1] which in fact is the source for most of the above descriptions in the literature.

B.1 Dynkin Diagrams

Let us consider a continuous group G of dimension d. Then there are d generators which satisfy the Lie algebra for the group. Let there be l commuting generators H_i ($i = 1, \cdots, l$) forming the Cartan subalgebra. There shall be even number of generators left, E_α, such that

$$[H_i, E_\alpha] = \alpha_i E_\alpha. \qquad (\text{B.1})$$

In equation (B.1) α_i are some of the structure constants of the group in the Cartan Weyl basis as above. The vector $\alpha = (\alpha_1, \alpha_2, \cdots \alpha_l)$ is called a root vector in the l-dimensional Euclidean root space. We can take the basis such that if α is a root, so is $-\alpha$. E_α thus is like a step up or step down operator of angular momentum. The other elements of the structure constants are given by the commutators

$$[E_\alpha, E_{-\alpha}] = \alpha_i H_i,$$

$$[E_\alpha, E_\beta] = N_{\alpha\beta} E_{\alpha+\beta}.$$

where $N_{\alpha\beta} = 0$ if $\alpha + \beta$ is not a root.

With respect to any basis we say that α is a **positive root** if the first nonvanishing component of α is positive. Clearly half of the nonzero

B.2. Dynkin Labels

roots are positive. The l positive roots which are linearly independent are called simple roots.

The length and angle relations between the simple roots completely characterises a Lie algebra. This leads to the definition of a two dimensional diagram, called **Dynkin diagram**, which represents the above properties of the simple roots and therefore characterises the group [2]. Each simple root is denoted by a dot "◯" on a diagram. For any group, the simple roots may have at best two different lengths. The longer roots are denoted by open dots (circles), and the shorter roots by filled in circles. The angle between two simple roots is denoted by a line connecting them. No line means that the angle is 90°, one line means that the angle is 120°, two lines means it is 135°, and, three lines means it is 150°. We note that the Dynkin diagram for the $SU(3)$ group is ◯——◯, indicating that it is group of rank two and has two simple roots of equal length with the angle between the roots being 120°. For a more complicated group, the exceptional group E_6 of rank six, the Dynkin diagram is given as

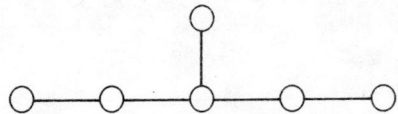

B.2 Dynkin Labels

The simple roots are not mutually orthogonal. The matrix which keeps track of their nonorthogonality, the **Cartan matrix**, is given as

$$A_{ij} = 2(\alpha_i, \alpha_j)/(\alpha_j, \alpha_j). \qquad (\text{B.2})$$

In the above, α_i *are roots, and not components of roots*. For any representation of the group, we can obtain different "states" or vectors by adding or subtracting the root vectors corresponding to the ladder operators E_α. Thus such states will have **weight vectors** λ the components of which are given by the eigenvalues of H_i and which are also vectors in the root space of dimension l as above. Let us express a weight vector Λ with coefficients $\bar{\lambda}_i$ in root space as

$$\Lambda = \sum_i \frac{2\bar{\lambda}_i}{(\alpha_i, \alpha_i)} \alpha_i. \qquad (\text{B.3})$$

Then the numbers $(\bar{\lambda}_1, \cdots \bar{\lambda}_l)$ represent the weight vector Λ in the **dual basis**. The **Dynkin labels** or **components** $(a_1, \cdots, a_l)$ of the weight vector Λ are defined to be

$$a_i = \frac{2(\Lambda, \alpha_i)}{(\alpha_i, \alpha_i)}$$

$$= \sum_j \frac{2\bar{\lambda}_j}{(\alpha_j, \alpha_j)} A_{ji}. \qquad (\text{B.4})$$

The selection of the above components or labels is worthwhile because of the following theorem, which we state without proof.

Theorem: *For any weight or root, the Dynkin labels as in equation (B.4) are integers.*

This makes the algebra of dealing with group representations reasonably simple.

When there are two weight vectors Λ and Λ', we can obtain the scalar product between them through the relation

$$(\Lambda, \Lambda') = \sum_i \bar{\lambda}_i a'_i = (\Lambda', \Lambda) = \sum_{ij} a'_i G_{ij} a_j, \qquad (\text{B.5})$$

where

$$G_{ij} = (A^{-1})_{ij}(\alpha_j, \alpha_j)/2. \qquad (\text{B.6})$$

In equation (B.6) we may note that the components in the dual basis of one weight vector multiplied by the Dynkin labels of the other gives the scalar product for the two weight vectors.

A linear combination of the commuting generators H_i defines an **axis** in root space. Thus, if Q is such a vector, a state $|\Lambda>$ with weight Λ is an eigenstate of Q, with eigenvalue $Q(\Lambda) = (Q, \Lambda)$. It is convenient to take Q in the dual basis $(\bar{q}_1, \cdots \bar{q}_l)$ so that we may write the above eigenvalue as

$$Q(\Lambda) = \sum_i \bar{q}_i a_i. \qquad (\text{B.7})$$

The above equation is very important regarding symmetry breaking patterns as noted e.g. in Chapter 12.

B.3 Representations

For a given group (i) we have to identify the different irreducible representations, (ii) obtain the weight vectors for a given representation, and, (iii) obtain automatically the dimension of each irreducible representation.

B.3. Representations

We note that any representation can be uniquely defined by one of the weights corresponding to that representation. The statement is made more precise by the following:

Theorem: *For each and every irreducible representation, we can uniquely identify a weight* $\alpha = (\alpha_1, \alpha_2, \cdots, \alpha_l)$ *such that each* $\alpha_i \geq 0$.

Such a weight vector is known as the **highest weight** for the representation. As noted, the Dynkin labels above are always integers. The **highest root** is the highest weight of the adjoint representation. *The adjoint representation of a group is the representation which has the same dimension as the dimension of the group,* and in fact the representation can be described by the matrices of the structure constants of the Lie group as may be verified by using Jacobi identity. The root vectors are weight vectors in the adjoint representation.

One can obtain the weight vectors of a representation by subtracting simple roots from the highest weight. We may trivially recognise the situation by considering the analogy with SU(3) flavor group. Here, starting from the highest weight, we can construct any other vector by repeated applications of the step down operators $I_-(u \to d)$ and $U_-(d \to s)$. The two operators I_+ and U_+ correspond to the simple roots and V_+ is a linear combination of the two. The beauty of the statement as above is that even when the group is much more complicated, a similar procedure works. E.g. E_6 group has thirty-six positive ladder operators, out of which, as above, we can always use six for subtraction as a step down operation, the repeated application of which will generate all the weights for a given representation by starting from the highest weight.

We note that the states with zero weight for a given representation shall be generally degenerate since this does not uniquely specify the state. However, all nonzero weight vectors give uniquely defined states [1]. Vectors in weight space may be distinguished from states in Hilbert space which they label. If the application of the above procedure gives a zero length vector in Hilbert space, such a state as usual is not admissible as a state for any representation since it is a zero vector in Hilbert space.

The **level** of a weight is the number of steps for such subtractions which is fixed for any weight although different simple roots in a different sequence may be used to reach it. Let $\Lambda = (a_1, \cdots, a_l)$ be the highest weight of a representation. The **level vector** is given in the dual basis by $\bar{R}_i = 2 \sum_j (A^{-1})_{ij}$, and with this, the **height** of a representation as given by the highest weight Λ is defined as $T(\Lambda) = (R, \Lambda)$. Then, the level of a weight with Dynkin labels $a' = (a'_1, \cdots, a'_l)$ is given as $\frac{1}{2}(T(\Lambda) - (R, a'))$. There are some symmetry properties of weights in root space; we shall not enter into them, since the above will be broadly adequate to obtain the different weight vectors for a representaion.

The dimension of a representation is given as

$$N(\Lambda) = \prod (\Lambda + \delta, \alpha)/(\delta, \alpha), \qquad (\text{B.8})$$

where $\delta = (1, 1, \cdots, 1)$ in Dynkin basis and the product above is to be taken over all the positive roots α.

There is the usual second order **Casimir invariant** C constructed from the generators which commutes with all the generators, given as

$$C = H_i G_{ij} H_j + \sum E_\alpha E_{-\alpha},$$

where summation over all roots α is to be taken. The eigenvalue of this invariant for a given representation is

$$C(\Lambda) = (\Lambda, \Lambda + 2\delta). \qquad (\text{B.9})$$

Another quantity, called the index $l(\Lambda)$ of a representation Λ is

$$l(\Lambda) = \frac{N(\Lambda)C(\Lambda)}{N(adj)}, \qquad (\text{B.10})$$

where $N(adj) = d$ is the dimension of the adjoint representation. $l(\Lambda)$ is always an integer. One relevant application of this is the one loop contribution to $\beta(g)$ for the scale dependance of the running coupling constant in renormalisation group equations. With $\beta(g) = \mu \frac{dg}{d\mu}$, we have in fact

$$\beta(g) = -\frac{1}{16\pi^2} \Big[\frac{11}{3} c(\text{vectors}) -$$

$$- \frac{2}{3} c(\text{Majorana spinors}) - \frac{1}{3} c(\text{complex scalars}) \Big], \qquad (\text{B.11})$$

where $c(..)$ is the index of the corresponding representation. This result has been utilised in Chapter 10 to evaluate the contributions from supersymmetric multiplets in renormalisation group equations.

If we have two representations R_1 and R_2 of the group G, we can consider the tensor product of these representatins and obtain

$$R_1 \otimes R_2 = \sum_i R_i, \qquad (\text{B.12})$$

where the right hand side consists of the direct sums of the irreducible representations of G, some of which may be repeated. This will be often needed since we shall have products of multiplets finally giving a scalar or a singlet for the group G. Further, let us have a subgroup H of G.

B.4. Symmetry Breaking

We can then consider the generators of G to consist of the generators of H and some more generators. An irreducible representation of G will then become sum of some irreducible representations H, which will depend on the way H is embedded in G. We shall have occasions to consider this when we consider symmetry breaking.

B.4 Symmetry Breaking

For the group G, let us assume that a scalar multiplet ϕ belonging to the representation r attains a nonzero vacuum expectation value, such that $\phi_0(r, \lambda)$ for some weights λ are nonzero constants. Now, the kinetic term for the scalar fields is given as $(D_\mu \phi)^\dagger (D^\mu \phi)$. Here $D_\mu = \partial_\mu - ig A_\mu$, with $A_\mu = A^a{}_\mu t_a$ and t'_as as matrices for the generators in the representation r. From the above we easily obtain that the mass term for the vector mesons is given as $(M^2)_{ab} A^a{}_\mu A^{b\mu}$, with the mass-squared matrix being

$$(M^2)_{ab} = g^2 \sum_\lambda (t_a \phi_0(r,\lambda)^\dagger)(t_b \phi_0(r,\lambda)). \qquad (\text{B.13})$$

The subgroup H of G defined by the zero eigenstates of the above mass matrix becomes the residual symmetry group after symmetry breaking. Alternatively, we may say that the maximal subgroup which keeps the vectors $\phi_0(r, \lambda)$ unchanged is the residual symmetry group.

If X_a are the generators of the group, we note that the scalar field is a tensor operator, such that we have

$$(t_a)_{ij} \phi_j = [X_a, \phi_i].$$

Hence an alternative form of the mass-square matrix is

$$(M^2)_{ab} = -g^2 \sum_\lambda Tr\{[X_{-a}, \phi_0(\bar{r}, -\lambda)][X_a, \phi_0(r, \lambda)]\}. \qquad (\text{B.14})$$

The above formula is more useful when we can express the Higgs multiplet in terms of the generators. In fact, let us have r as the adjoint representation and ϕ_0 as nonzero only for zero weight. We can then write

$$\phi_0(r, \lambda = 0) = \bar{c}_i H_i, \qquad (\text{B.15})$$

where, parallel to equation (B.7), $(\bar{c}_1, \cdots, \bar{c}_l)$ defines an axis in root space in dual basis. The axis here shall be called as the **symmetry breaking axis**. The mass-square matrix then becomes diagonal. With $a = (a_1, \cdots a_l)$ being the root in Dynkin basis corresponding to the generator X_a, using equation (B.1), we in fact obtain that

$$(M^2)_{ab} = g^2 \delta_{ab}(a,c)^2. \tag{B.16}$$

Thus with the Higgs particles in the adjoint representation, the mass becomes proportional to the projection of root vectors onto the symmetry breaking axis $\bar{c}$ as in equation (B.15).

B.5 E_6 Symmetry Group

As discussed in Chapter 12, E_6 group is very important in the context of compactification of ten dimensional supergravity to four dimensions. Let us consider the above symmetry group with a simple consequence of equation (B.16). For E_6 group, let us take the general symmetry breaking axis in dual basis as

$$\bar{c} = (a, b, c, d, e, f). \tag{B.17}$$

We shall examine what should be the condition on the above vector so that $G_{321} \equiv SU(3)_C \times SU(2)_L \times U(1)_Y \equiv G_{std}$ remains unbroken. For this purpose we have to embed G_{std} in E_6. We do so with the following conventions. For the adjoint representation, the roots of E_6 will specify the states. Half of the nonzero roots are given in Table 1. For this representation the highest weight is $(0,0,0,0,0,1)$. In our embedding we identify this to belong in $SU(3)_C$ space. In addition, we may take another root (or weight) as $(0,1,0,0,-1,0)$ in $SU(3)_C$. All other nonzero roots for $SU(3)_C$ are linearly dependant on the above two. Now, from equation (B.16), the requirement that the gauge bosons corresponding to $SU(3)_C$ are massless becomes equivalent to the vanishing of the projections of the above roots on the symmetry axis. This gives rise to two equations which give the condition that

$$\bar{c} = (a, b, c, d, b, 0). \tag{B.18}$$

The above embedding can now be completed using Table 1 with the assignment for a nonzero root for $SU(2)_L$ as $(1,0,0,0,1,-1)$. The symmetry breaking axis shall also be orthogonal to this for G_{std} to remain unbroken. We thus obtain the constraint for the symmetry breaking axis which does not break Salam Weinberg symmetry to be given as

$$\bar{c} = (a, -a, c, d, -a, 0), \tag{B.19}$$

where a, c and d remain arbitrary. We may mention that looking up Dynkin labels from a table as above enabled us [1] to resolve a very complicated problem with comparative ease.

B.5. E_6 Symmetry Group

We note that from equation (B.16) the vector mesons corresponding to the Cartan subalgebra always remain massless. Hence this mode of symmetry breaking will always keep the rank of the group unaltered even after symmetry breaking with a few $U(1)$ factors hanging around. E.g. with equation (B.19) in general the broken symmetry would be

$$SU(3)_C \times SU(2)_L \times U(1)_Y \times U(1) \times U(1). \qquad (B.20)$$

In Chapter 12 the above mechanism is utilised for symmetry breaking for multiply connected spaces through Wilson loops. The manifold structure as described in Appendix A and the group structure as discussed here become closely linked giving rise to a very unusual type of symmetry breaking. We shall be particularly utilising equation (B.7) to identify the quantum numbers for the different states as well as equations like (B.19) and Table 1 to obtain constraints due to anticipated symmetry properties after symmetry breaking. For this purpose we shall need some further details of E_6 group which we shall give here. We have identified a set of simple roots in Table 1 along with half the nozero roots of the E_6 group as these will enable us to obtain the weights for other representations of the group when we know the highest weight. We shall be particularly interested in the 27-plet which will describe the matter sector.

B.5.1 Embedding of G_{321} in E_6

We note that after E_6 symmetry breaking, the Salam Weinberg group G_{321} must be good. However, the way this embedding can be done is not unique. We shall consider some of the options available here. In the last paragraph we have already started the topic where we have oriented our discussions for a specific embedding. The relevant feature of any embedding will be the structure of the 27-plet which will have to give the quantum numbers of quarks and leptons correctly. Thus for convenience we have noted the weights of the 27-plet of E_6 in Table 2 along with the way they can be obtained from the state of highest weight for the representation by repeatedly subtracting the simple roots of Table 1 as an illustration. The "spindle shape" of the diagram, true for any such diagram for any group [1], may be noted. These weights will enable us to know the quantum numbers through equation (B.7).

Let us identify three diagonal generators U_t, U_r and U_w of E_6 through the following chain :

$$E_6 \supset SO(10) \times U_t(1) \supset SU(5) \times U_r(1) \times U_t(1)$$

$$\supset SU(3)_C \times SU(2)_L \times U_w(1) \times U_r(1) \times U_t(1). \qquad (B.21)$$

We note that in the above we have not identified the weak hypercharge Y, which could be any combination of the charges Q_w, Q_r and Q_t that shall be consistent with the quantum numbers of the quarks and leptons. We shall now consider some of the multiple choices as may be available to us.

Let us first note that the above operators in Dynkin basis and in dual basis are respecitively given as [1]

$$Q_t = (3, -3, 0, 3, -3, 0); \quad \bar{Q}_t = (1, -1, 0, 1, -1, 0);$$

$$Q_r = (-3, -1, 4, 1, -1, -4); \quad \bar{Q}_r = (-1, 1, 4, 3, 1, 0);$$

$$Q_w = \tfrac{1}{3}(3, -4, 6, -6, 1, -1); \quad \bar{Q}_w = \tfrac{1}{3}(1, -1, 1, -3, -1, 0). \quad (\text{B.22})$$

In the above we have by equation (B.16) e.g that the symmetry breaking direction of $\bar{Q}_t$ will leave the $SU(6)$ gauge bosons massless, and similarly the symmetry breaking direction of $\bar{Q}_r$ will leave the gauge bosons of $SU(5)$ as massless. With the embedding of $SU(2)_L$ as above, we also have $I_{3w} = \tfrac{1}{2}(1, 0, 0, 0, 1, -1)$ and $\bar{I}_{3w} = \tfrac{1}{2}(1, 1, 1, 1, 1, 0)$ in Dynkin basis and in dual basis respectively.

The simplest embeddding of G_{321} is through $SU(5)$ with the usual identification of the quarks and leptons. In that case, we have $Y = Q_w$ and the electromagnetic charge Q_{em} is given as

$$Q_{em} = I_{3w} + \frac{1}{2}Y. \quad (\text{B.23a})$$

We shall name the embedding with the above identification of charge as the conventional embedding. The multiplet structure for the 27-plet is then given as in Table 3.

We next give another type of embedding with

$$Y = \frac{1}{10}(Q_w + 2Q_r)$$

and the identification for electromagnetic charge as

$$\bar{Q}_{em} = -I_{3w} - \frac{1}{2}Y. \quad (\text{B.23b})$$

Clearly in equation (B.23b) the neutrino and the electron, as well as the u and d quarks will flip their positions as $SU(2)_L$ doublets. Hence, the symmetries associated with this identification of charge and multiplet structure shall be known as "flipped" symmetries [3]. We may thus have flipped $SU(5)$ or $SU(6)$ symmetries in a chain for symmetry breaking when we start from E_6.

We shall consider some of these choices in Chapter 12.

Table 1

Nonzero E_6 roots (half of them)

α_i ($i = 1, \cdots, 6$) are simple roots

Color SU(3) roots
(0,0,0,0,0,1)
(0,1,0,0,-1,0)
(0,-1,0,0,1,1)
Left handed SU(3) roots
(1,0,0,0,1,-1)
(-1,1,0,0,1,-1)
(2,-1,0,0,0,0) = α_1
Right handed SU(3) roots
(0,-1,1,1,-1,-1)
(0,0,-1,2,-1,0) = α_4
(0,-1,2,-1,0,-1) = α_3
SU(5) antileptodiquarks
(1,-1,1,-1,1,0)
(1,0,1,-1,0,-1)
(1,-1,1,-1,1,-1)
(0,-1,1,-1,0,1)
(0,0,1,-1,-1,0)
(0,-1,1,-1,0,0)
SO(10)/SU(5) leptoquarks
(0,0,1,0,0,-1)

(0,-1,1,0,1,-1)
(0,0,-1,0,0,2) = α_6
(-1,0,1,0,-1,0)
(-1,-1,1,0,0,0)
(-1,0,1,0,-1,-1)
(-1,0,0,1,0,0)
(-1,1,0,1,-1,-1)
(-1,0,0,1.0,-1)
E_6/SO(10) leptoquarks
(0,1,0,-1,1,0)
(0,0,0,-1,2,0) = α_5
(0,1,0,-1,1,-1)
(-1,1,0,-1.0,1)
(-1,0,0,-1,1,1)
(-1,1,0,-1,0,0)
(-1,0,1,-1,1,0)
(-1,1,1,-1,0,-1)
(-1,0,1,-1,1,-1)
(-1,1,-1,0,1,1)
(-1,2,-1,0,0,0) = α_2
(-1,1,-1,0,1,0)

Table 2
Weights of 27-plet representation of E_6

$$(100000)$$
$$\alpha_1 \downarrow$$
$$(-110000)$$
$$\alpha_2 \downarrow$$
$$(0-11000)$$
$$\alpha_3 \downarrow$$
$$(00-1101)$$
$$\alpha_4 \swarrow \quad \searrow \alpha_6$$

$(000-111)$ $\qquad\qquad$ $(00010-1)$
$\alpha_5 \downarrow \quad \alpha_6 \searrow \quad \alpha_4 \downarrow$
$(0000-11)$ $\qquad\qquad$ $(001-11-1)$
$\alpha_6 \downarrow \quad \swarrow \alpha_5 \quad \alpha_3 \downarrow$
$(0010-1-1)$ $\qquad\qquad$ $(01-1010)$
$\alpha_3 \downarrow \quad \alpha_4 \swarrow \quad \alpha_2 \downarrow$
$(01-11-10)$ $\qquad\qquad$ $(1-10010)$
$\alpha_4 \swarrow \quad \alpha_2 \searrow \qquad \alpha_5 \swarrow \quad \alpha_1 \searrow$

$(010-100) \qquad (1-101-10) \qquad (-100010)$
$\alpha_2 \searrow \qquad \alpha_4 \swarrow \quad \searrow \alpha_1 \qquad \swarrow \alpha_5$
$\qquad (1-11-100) \qquad (-1001-10)$
$\qquad \alpha_3 \downarrow \quad \alpha_1 \searrow \quad \alpha_4 \downarrow$
$\qquad (10-1001) \qquad (-101-100)$
$\qquad \alpha_6 \downarrow \quad \alpha_1 \searrow \quad \alpha_3 \downarrow$
$\qquad (10000-1) \qquad (-11-1001)$
$\qquad \alpha_1 \downarrow \quad \alpha_6 \swarrow \quad \alpha_2 \downarrow$
$\qquad (-11000-1) \qquad (0-10001)$
$\qquad \alpha_2 \searrow \qquad \swarrow \alpha_6$

$$(0-1100-1)$$
$$\alpha_3 \downarrow$$
$$(00-1100)$$
$$\alpha_4 \downarrow$$
$$(000-110)$$
$$\alpha_5 \downarrow$$
$$(0000-10)$$

B.5. E_6 Symmetry Group

Table 3
Multiplet structure of 27-plet

E_6 weight	Color	Name	Q_{em}	I_{3w}	Q_t	SU_5	$SO(10)$
00010-1	00	ν	0	1/2	1	5*	16
-1001-10	00	e	-1	-1/2	1	5*	16
1-11-100	00	e^c	1	0	1	10	16
10-1001	00	ν^c	0	0	1	1	16
001-11-1	00	H^+	1	1/2	-2	5	10
-101-100	00	H^0	0	-1/2	-2	5	10
01-1010	00	$\tilde{H}^0$	0	1/2	-2	5*	10
-11-1001	00	$\tilde{H}^-$	-1	-1/2	-2	5*	10
1-101-10	00	N	0	0	4	1	1
100000	10	u	2/3	1/2	1	10	16
1-10010	-11	u	2/3	1/2	1	10	16
10000-1	0-1	u	2/3	1/2	1	10	16
0000-11	10	d	-1/3	-1/2	1	10	16
0-10001	-11	d	-1/3	-1/2	1	10	16
0000-10	0-1	d	-1/3	-1/2	1	10	16
-110000	10	D	-1/3	0	-2	5	10
-100010	-11	D	-1/3	0	-2	5	10
-11000-1	0-1	D	-1/3	0	-2	5	10
000-111	01	$\tilde{D}$	1/3	0	-2	5*	10
010-100	1-1	$\tilde{D}$	1/3	0	-2	5*	10
000-110	-10	$\tilde{D}$	1/3	0	-2	5*	10
0-11000	01	d^c	1/3	0	1	5*	16
0010-1-1	1-1	d^c	1/3	0	1	5*	16
0-1100-1	-10	d^c	1/3	0	1	5*	16
00-1101	01	u^c	-2/3	0	1	10	16
01-11-10	1-1	u^c	-2/3	0	1	10	16
00-1100	-10	u^c	-2/3	0	1	10	16

REFERENCES

1. R. Slansky, Phys. Rep. 79, 1 (1981).

2. E.B. Dynkin, Amer. Math. Soc. Trans. Ser.2.6, 111 and 245 (1975).

3. S.M. Barr, Phys. Lett. 112B, 219 (1982).

Index

Abelian gauge group 38
Action integral 70,91,94
Action principle, Schwinger 10
Anomaly cancellation 58,132
Atlas 207
Auxiliary field 25,27,40,78
Auxiliary variable 15

B meson 50
Betti-Hodge numbers 152,209,213,220
Bianchi identities 84
Bino 50
Bogoliubov transformation 187

CHSW manifold 215,216
CICY space 215
CP^3 151,216
CP^4 150,215
CP^n 213
Cabbibo rotation 49
Calabi Yau space 150,155,157,211
Cartan matrix 225
Cartan subalgebra 156,224
Chart 207
Casimir invariant 110,228
Chern class 150,218
Chern Simon term 132
Chiral density 91
Chiral multiplets 26,40,88
Chiral scalar curvature 94
Chiral superfield 25,28,38
Christoffel symbol 83
Closed forms 209

Compactification 129,139,157
Compactification scale 164
Complex deformation 220
Complex manifolds 207
Complex structure 150,208
Component fields 24
Condensate 120,171,198
Connection 83,210
Coset space 172
Cosmological constant 35,115
Covariant derivative 9,40,85,95,210
Curvature 83,210

D-type term 29
D-type symmetry breaking 54
Differential forms 81,209
Dilaton 129,137
De Sitters space 149
Dirac spinors 20
Dual basis 225
Dual spinor model 128
Dynkin diagram 224
Dynkin labels 225,226

E_6 group 65,143,230,233,234,235
E_8 group 65,132
Einstein vector 83
Electroweak theory 46
Embedding in E_6 231
Euler characteristic 218
Euler number 151,219
Exact forms 209
Exterior derivative 82

F-type term 29,31
F-type symmetry breaking 54
Fayet-Illiopoulos mechanism 54,61
Feynman rule 75
Flat superspace 85
Flipped symmetry 170,232
Flux breaking 153,154
Flux quantisation 222
Functional differentiation 72

Gauge connection 142,143
Gauge fixation 27
Gauge superfield 38
Gauge transformation 38
Gaugino field 38,40
Gellmann charge 47,49,232
General coordinate transformation 86
Generators 11,23
Gluino 50
Gluons 50
Goldstone mode 56,191
Grand unification 48,64,104,107
Grassmann variable 7,69
Gravitino 50,61,143
Graviton 50
Gravity multiplet 88
Greens function 73,75
Group representation 226

Harmonic forms 217
Heavy modes 140
Hermitian manifold 208
Heterotic string 133
Hexagonal diagram 132
Hidden sector 104
Hierarchy generation 113
Higgs particles 50,60,118
Higgsino 50
Hodge decomposition 218
Hodge diamond 213
Holonomy, SU(3) 144,152
Hosotani mechanism 153

Index 228
Integration, Grassmannian 69
Intermediate scales 112,165

Kahler manifold 149,210
Kahler metric 99
Kahler potential 97,210
Kaluza-Klein theory 128,138,172
Kobayashi-Maskawa rotation 49,58

Left-right symmetry 113,200
Lie algebra 224
Light cone gauge 134
Local coordinates 83
Local supersymmetry 84
Lorentz Chern-Simon term 132
Lorentz vector 83

Majorana spinor 20
Manifold 207
Matrix models 200
Maximally symmetric space 148
Metacolor 196
Multiply connected space 151

$N \to \infty$ approximation 204
Neutral currents 109
Neutrino 50,117
Neveu Schwarz model 134
Niejenhuis tensor 208
Nonrenormalization theorem 76

O'Raifeartaigh mechanism 53

Particle multiplets 50,163,235
Partition function 72,74
Pati Salam symmetry 112
Phase transition 55,61,183
Planck scale 50,66,105,107,126,144
Polonyi singlet 108
Polonyi superpotential 103,108
Potential 46
Preonic models 195
Projective space 213
Propagator 73

INDEX

Proton decay 56,111

Quantum gravity 126
Quantum phase transition 61,183
Quark 50

R-charge 67
R-parity 66
R-symmetry 66
Radiative symmetry breaking 122
Ramond model 134
Ramond vacuum 136
Rarita-Schwinger field 129
Ricci 2-form 210
Ricci tensor 87,211
Root vector 224
Running coupling constant 111,228

S-matrix 77
$SL(2,C)$ 17
$SO(10)$ 65,107
$SO(32)$ 133
$SU(5)$ 57,64,104
$SU(6) \times U(1)$ 169
SUSY 1+1 dimension 175
SUSY finite temperature 186
Salam Weinberg theory 49
Scalar curvature 87
Second hierarchy problem 117
Soft symmetry breaking 59,60
Spin connection 142,143
Spinor field strength 41
Standard model 48,118,231
Super coordinate 7
Super Higgs mechanism 105
Super gauge transformation 86
Super graphs 80
Superfields 25,26,33,177
Supergravity transformation 89
Superheavy Higgs particles 118
Superpotential 9, 31
Superpropagator 73
Superspace 7,23,69,81,176
Superstrings 133,152

Supersymmetric charge 14,180
Symmetry breaking 50,107,145
Symmetry breaking axis 229

Technicolor 171
Thermofield method 186
Tian Yau manifold 216
Top quark 125
Torsion 84,85

Vacuum degeneracy 105,134
Vacuum destabilisation 64,192
Variational method 186,191
Vector superfield 26,32
Vector multiplet 27,38
Vielbein 83,129
Vielbein superfield 83,85

W meson 50
Ward identity 197
Weight vector 225
Wess-Zumino gauge 27, 39
Weyl rescaling 96
Weyl spinor 20
Wilson loop 153
Wino 50

Yang-Mills 42,45,132
Yukawa coupling 30,101,123,160

Z_2 symmetry 154,184
Z_3 symmetry 158
Z_5 symmetry 156

EPILOGUE

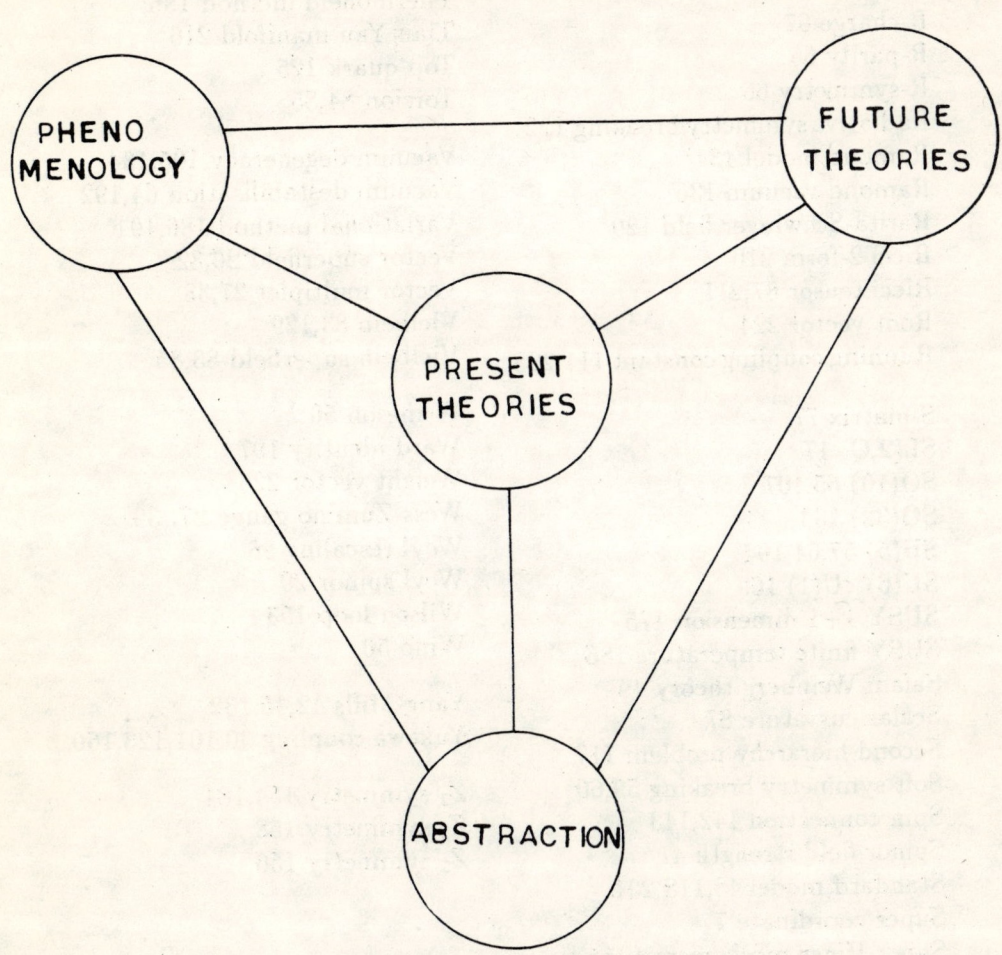

Love Thy Neighbour